ARTHROPOD NATURAL ENEMIES IN ARABLE LAND

II

Proceedings of the Second EU Workshop on Enhancement, Dispersal and Population Dynamics of Beneficial Insects in Integrated Agroecosystems: "Estimating survival and reproduction of beneficial predators and parasitoids in relation to food availability and quality of the habitat", held at Wageningen International Conference Centre, The Netherlands, 1-3 December 1994

ARTHROPOD NATURAL ENEMIES IN ARABLE LAND

II

Survival, Reproduction and Enhancement

Edited by Kees Booij and Loes den Nijs

ACTA JUTLANDICA LXXI:2
Natural Science Series 10

AARHUS UNIVERSITY PRESS

Printed in England by the Alden Press
ISBN 87 7288 672 2
ISSN 0065 1354 (Acta Jutlandica)
ISSN 0105 6824 (Natural Science Series)

AARHUS UNIVERSITY PRESS
University of Aarhus
DK-8000 Aarhus C
Fax (+45) 8619 8433

73 Lime Walk
Headington, Oxford OX3 7AD
Fax (+44) 1865 750 079

Box 511
Oakville, Conn. 06779
Fax (+1) 203 945 9468

Foreword

This book is the second in a series of three volumes that will form the proceedings of workshops funded as a Concerted Action by the European Union. All three workshops deal with the ecology of arthropod predators and parasitoids and the ways to utilize their role in limiting pest insects in European agriculture. Recent scientific developments and methodological problems associated with studies of these beneficial insects formed the central theme of the workshop series. The common title of the concerted action and the three workshops is:

ENHANCEMENT, DISPERSAL AND POPULATION DYNAMICS OF BENEFICIAL PREDATORS AND PARASITOIDS IN INTEGRATED AGROECOSYSTEMS.

This book is one of the results of the second workshop which took place at Wageningen International Conference Centre, The Netherlands, from 1-3 December 1994. It followed an earlier workshop in Aarhus, Denmark in October 1993 and preceded the third workshop in Bristol, United Kingdom which was held in November 1995.

Each of the workshops had its own subject area and themes. Accordingly different groups of experts and enthusiastic scientists from all over Europe participated in the workshops. They presented papers, posters and contributed to the brainstorm and discussion sessions which formed an essential part of the workshops.

The themes of the three workshops respectively were:

1993 Aarhus — *Estimating population densities and dispersal rates of beneficial predators and parasitoids in agroecosystems.*

1994 Wageningen — *Estimating survival and reproduction of beneficial predators and parasitoids in relation to food availability and quality of the habitat.*

1995 Bristol — *Analysing and modelling of population dynamics of beneficial predators and parasitoids in agroecosystems.*

The background of the concerted action was a general feeling among researchers that progress in the work on beneficials was hampered by serious methodological problems in field ecology and the absence of a common framework to promote large scale research and cooperation between different research groups. Without doubt beneficial arthropods are important for future sustainable agriculture. However, our poor understanding of the ecology of many species is a major handicap in exploiting their potential for the control of harmful insects. For a full account on the arguments for organizing the workshops the reader is referred to the first volume of this series.

The fact that in agriculture beneficials and the people who study them have to cope with a very dynamic habitat where disturbance regimes are severe and where processes at the field and landscape scale are tightly linked runs as a thread through all the workshop discussions. Fortunately, most researchers consider this complexity as a challenge.

An integrated management of our crops and farms and future changes in land use will have major impacts on the dynamics and functioning of the natural enemies of insect pests. We think that understanding these impacts is necessary in providing a better ecological base for sustainable production.

Whereas measuring population densities and movement was the central theme of the first workshop, the second workshop focused on the factors underlying changes in densities, being reproduction, survival and the ways densities can be manipulated by changing the habitat.

Presentations of research and discussion sessions were of equal importance in the workshop. As most manuscripts were written after the workshop much of the discussion items have been incorporated in this book. Apart from that, the leaders of the discussions have summarized or even reviewed central issues in separate papers. We hope that the combination of research results, the many aspects that have been discussed, and the numerous literature references in this book provides the reader with a firm background for the study of beneficials.

We are particularly grateful to all the workshop participants for their contribution and for their mutual reviewing of the manuscripts.
We would also like to thank Clasien Lock, Jan Noorlander and Richard Daamen for their help in organizing the second workshop and realizing this book.

Wageningen, July 1996

Kees Booij Loes den Nijs

Contents

SURVIVAL

DISCUSSION PAPER

REPRODUCTION

DISCUSSION PAPER

ENHANCEMENT

DISCUSSION PAPER

SURVIVAL

The influence of environmental factors and food on life cycle, ageing and survival of some carabid beetles

Th.S. van Dijk

Biological Station, Centre for Soil Ecology,
Wageningen Agricultural University,
Kampsweg 27, 9418 PD Wijster, The Netherlands

Abstract
Reproduction and survival of carabids are determined by a complex of variable factors (e.g. weather factors, age of the individuals, individual variability and quantity or quality of food). The influences of temperature, substrate moisture and the quality or quantity of food on reproduction, survival and growth of two carabid beetles, *Calathus melanocephalus* and *Pterostichus versicolor*, are discussed.

The period of egg laying and the mean number of eggs laid per female per week are positive correlated with temperature and food supply.

A positive relationship exists between the quantity and quality of food ingested and the size of egg production.

Temperature and substrate moisture strongly influence the mortality of all developmental stages.

The length of the larval growth and adult body size are influenced by temperature and food supply. No inverse relationship (trade-off) occurs between reproductive output (manipulated by different temperature regimes and an excess of food or by different levels of food at the same temperature) and survival in the field until the next breeding season.

It is discussed to what extent the results explain the differences between fluctuation patterns of populations and whether or not carabids usually experience a more or less continuous shortage of suitable food, especially that which contains enough nitrogen.

Key words: *Calathus melanocephalus, Pterostichus versicolor*, quantity and quality of food, reproduction, population dynamics

---

Arthropod natural enemies in arable land · II *Survival, reproduction and enhancement*
C.J.H. Booij & L.J.M.F. den Nijs (eds.). *Acta Jutlandica* vol. 71:2 1996, pp. 11-24
 ISBN 87 7288 672 2

Introduction

The numbers of individuals in carabid populations depend on the number of offspring per female, the survival of adults from one breeding season to the next and the exchange of individuals between populations. Here attention will be given to factors influencing reproduction and survival. In all life stages, eggs, three larval stages and a pupal stage, mortalities occur. Moreover the numbers of adults that succeed to survive one or more years are highly variable (Baars and Van Dijk 1984, Van Dijk 1982) and depend on the species concerned (Nelemans c.s. 1989).

A complex of factors e.g. weather factors, age of the individuals, individual variability, quantities or qualities of food, etc. influence reproduction and survival of carabids in the field (Van Dijk 1982, 1986a,b, Van Dijk and Den Boer 1992).

At first data will be presented about the influence of temperature, substrate moisture and food on the size of egg production and on the larval growth and/or survival of two carabid species: *Calathus melanocephalus* and *Pterostichus versicolor* (Van Dijk 1986a,b, Van Dijk and Den Boer 1992, Van Dijk 1994). Particular attention will be given to the influence of the quality of food (i.e. food with different percentages of nitrogen) on egg production and larval growth in *Bembidion tetracolum*.

Next results will be given from experiments conducted in the field with two carabid species to test whether it is true that an inverse relationship exists between the size of egg production, which is directly related to food supply and temperature, and the survival of adults from one breeding season to the next (Van Dijk 1979a, 1994). Several authors suggested such an inverse relationship between reproduction and survival which should act as a compensatory mechanism leading to population stability in Carabidae (Murdoch 1966a,b, Price 1984, Stearns 1992).

Finally it is discussed to what extent the results explain the differences between fluctuation patterns of populations and whether or not carabids usually experience a more or less continuous shortage of suitable food.

Material and Methods

Size of egg production

With regard to the relationship between the size of reproduction and temperature can be referred to in the discussion about the inverse relationship between the size of reproduction and adult survival from one breeding season to the next.

In the breeding season females of *P.versicolor* and *C. melanocephalus* were divided into three groups and kept with different quantities of food (blowfly larvae in *P. versicolor* and *Drosophila* larvae in *C. melanocephalus*) at the same temperature (for details cf. Van Dijk 1994).

To determine the effect of food quality six groups of ten individuals each of *B. tetracolum* were used in 1989. Each individual was kept separately in a Petri dish (Ø 6 cm) with moist peat dust and at changing temperatures: 18°c (12 hrs, at day) and 12°c (12 hrs, at night). Each group was fed either with eggs (0.53 mg) of *P. versicolor*: 1, 2 or 3 eggs per female per day or fed with blowfly larvae (0.85-1.54 mg): 2, 3 or 4 larvae per female per day. Every week the number of eggs laid were sampled and counted by means of a washing method (cf. Mols et al. 1981). This experiment lasted seven weeks. Results of the first three weeks and the last two weeks were discarded. The first three weeks were an adaptation period. The last two weeks were excluded because of a sharp decrease in oviposition in connection with the end of the reproduction period.

Carbon and nitrogen contents of the two kinds of food and of the laid eggs of *B. tetracolum* were determined by bomb calorimetry.

Larval development

Newly hatched first-instar larvae of *P. versicolor* (a day active beetle in spring and summer) and *C. melanocephalus* (a night active beetle in summer) were kept individually in small Petri dishes with moist ground peat dust. The larvae of *P. versicolor* were kept at 19°c, those of *C. melanocephalus* at 15.5°c. The larvae were divided into two groups: one group supplied with a small amount of food and a second group with an excess of food (blowfly larvae) (for the use of other constant temperatures and field temperatures see Van Dijk 1994).

A similar experiment was executed with *B. tetracolum.* At 15.5°c first-instar larvae were kept in small Petri dishes with moist filter paper. Two groups of ten larvae were made. One group was fed three times a week with eggs (± 0.53 mg) of *P. versicolor* (first-instar larva one egg and the two other stages two eggs per larva). The second group was fed three times a week with blowfly larvae (at first 1.2 mg per first-instar larva up to 6.4 mg per third-instar larva).

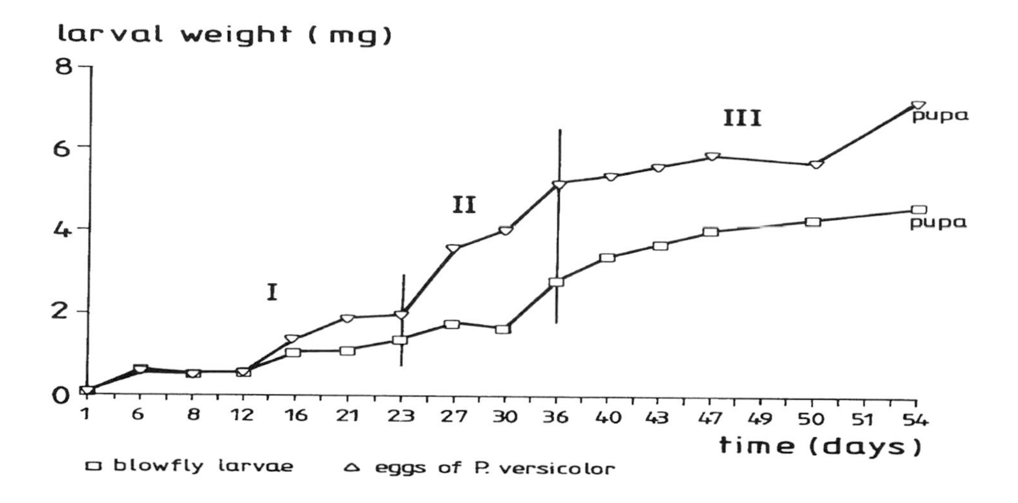

Fig. 1. *B. tetracolum*: The relationship between the amounts of different kinds of food consumed per female per day and the numbers of eggs laid per female per day. (The maximum consumption of each category of food is indicated by arrows).

Mortality of the developmental stages of P. versicolor and C. melanocephalus
Both species were reared at five constant temperatures and three levels of soil moisture and an excess of food (for details see Van Dijk and Den Boer 1992).

Inverse relationship between reproduction and survival
An inverse relationship between the size of reproduction and the adult survival from one breeding season to the next is often assumed.

If such an inverse relationship exists, the highest mortalities should be expected among females in the field during winter, with the highest reproduction effort in spring or summer. Hence in two ways the reproductive effort of *P. versicolor* and *C. melanocephalus* was manipulated. The first way was by keeping females in the laboratory at different temperatures and an excess of food (in *C. melanocephalus*). In this way we also got information about the relationship between the size of reproduction and temperature. The second way was by keeping females at one temperature and by feeding them with different quantities of food (blowfly larvae). At the end of the breeding season the beetles were marked individually and released inside enclosures in

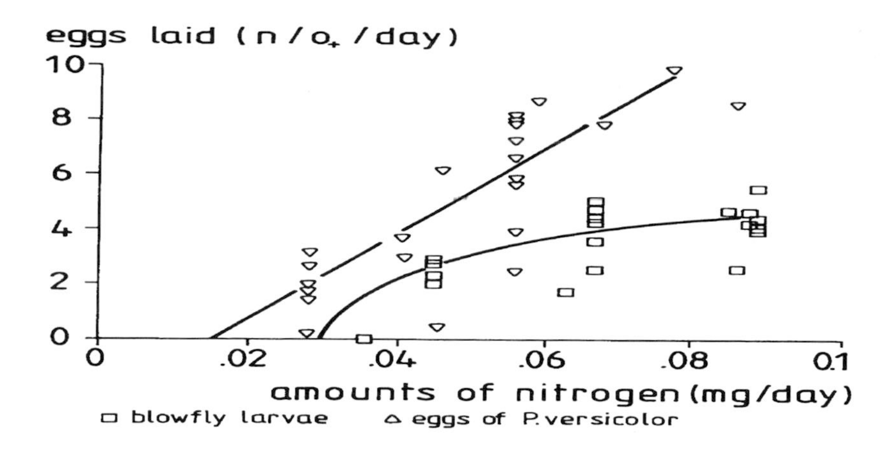

Fig. 2. *B. tetracolum*: The relationship between the amounts of nitrogen consumed per female per day and the numbers of eggs laid per female per day by using different kinds of food.

the field. During the next year from May until September the beetles were recaptured by means of pitfall traps from these enclosures to determine numbers lost (= winter mortality). Emigration of the enclosures could be ignored (for details see Van Dijk 1994).

Results

Size of egg production

In *C. melanocephalus* a significant positive correlation was found between temperature and the mean number of eggs laid per female per season ($r=0.99$, $p=0,02$, Table 4). For more data about this relationship for both species cf Van Dijk 1994).

A positive relationship existed between food supply and the mean number of eggs laid per female of *P. versicolor* or *C. melanocephalus* per season (Table 1) (Van Dijk 1986a, 1994).

The number of eggs laid per day by *B. tetracolum* was much higher when they were nourished with eggs of *P. versicolor* than when nourished with comparable or higher quantities of blowfly larvae (Fig. 1). Females from the

Table 1. Egg production (per season), body weights (mg): mean and SD (in italics) and winter mortality (- % of numbers of released marked beetles within enclosures in 1980 and not recaptured in the next year) of *P. versicolor* and *C. melanocephalus* in relation to food supply.

Amounts of food	No. re-leased	No. re-captured	% loss	Egg Production		Body weights			
						1980		1981	
P. versicolor									
4mg/day	20	19	5.0	7.5	*11.7*	54.8	*8.3*	59.5	*12.3*
8mg/day	20	20	0.0	96.5	*65.4*	65.9	*6.3*	60.2	*11.2*
20mg/day	20	17	15.0	106.6	*83.3*	63.6	*7.3*	60.4	*8.3*
C. melanocephalus									
1mg/day	19	9	52.6	50.9	*21.5*				
2mg/day	20	11	45.0	73.8	*54.5*	not estimated			
5mg/day	19	10	47.4	142.5	*86.0*				

groups with the highest quantities of food per day (three eggs of *P. versicolor* and four blowfly larvae per day respectively) did not succeed to eat all the food. Thus there was an excess of food in these categories and it was possible to estimate the maximum consumption of *B. tetracolum* within each category of food: 1.5 mg eggs/day and about 5 mg blowfly larvae per day (indicated by arrows in Fig. 1). Apparently the quality of the ingested food has an important influence on maximal quantities of food consumed per female per day. In Fig. 2 the different quantities of food consumed per day per female from each category of food is represented by the quantities of nitrogen in mg/day. The maximum amount of food consumed per female was about the same within the two different food groups: about 0.09 mg/day (Fig. 2). The relationship between food supply and egg production is rectilinear when eggs were presented as food and curvilinear with blowfly larvae as food. Apparently with blowfly larvae (even when present ad libitum) females did not succeed in getting enough proteins to increase their egg production in a rectilinear way. Moreover, the quality of food not only influenced the size of egg production but also the qualities of eggs (represented by the percentages of nitrogen and carbon of them) laid by *B. tetracolum* (Table 2).

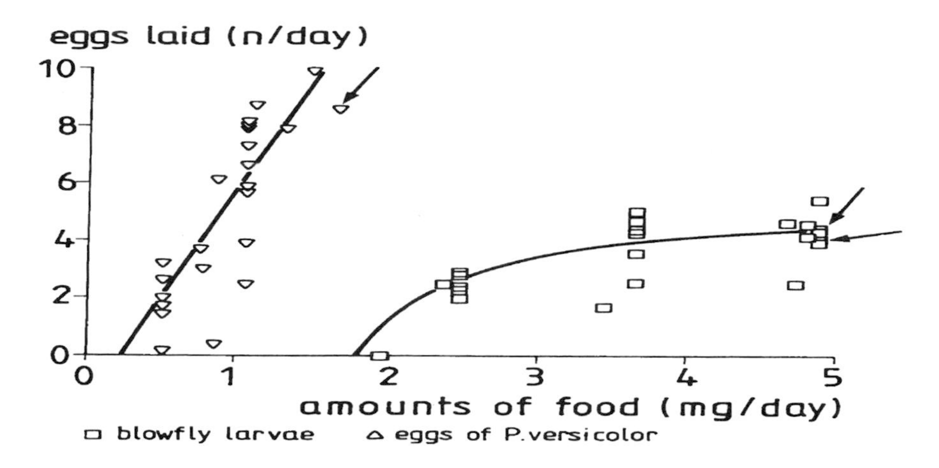

Fig. 3. *B. tetracolum*: Growth of larvae reared with two different kinds of food (I, II and III = the three larval stages).

Larval development

The development time from first larval stage until adult was significantly shorter and the adult weight higher with favourable than with less favourable food conditions at 19°c in *P. versicolor* and at 15.5°c in *C. melanocephalus* (Table 3). (For results with other constant temperatures and field conditions cf Van Dijk 1994).

In *B. tetracolum* the weights of the different larval stages were distinctly higher if they were fed with nitrogen-rich food (= eggs of carabids) than the weights of those fed with nitrogen-poor food (= blowfly larvae) (Fig. 3). However the lengths of the development period of each stage at the different food treatments was about the same.

Table 2. The percentages carbon and nitrogen present in the eggs laid by *B. tetracolum* when they were nourished with eggs of *P. versicolor* or with blowfly larvae.

Food	Carbon (%)	Nitrogen (%)
Eggs	45,9	9,6
Blowfly larvae	24,1	4,7

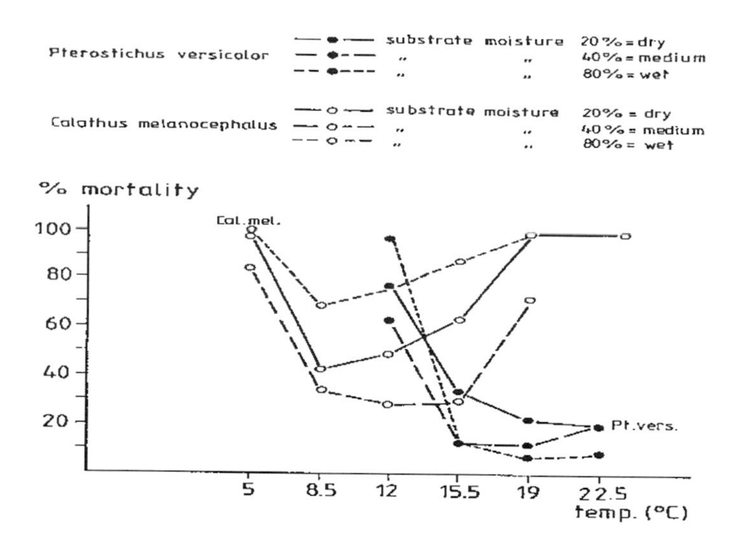

Fig. 4. *P. versicolor* and *C. melanocephalus*: Percentage mortality during development from egg to adult in relation to moisture and temperature.

Table 3. Length of development (days) from the first larval stage until adult and the weights (mg) of the adults of *P. versicolor* at 19°c and of *C. melanocephalus* at 15.5°c at a low and high level of food amounts (N = number of animals, X= mean and S = standard deviation).

	low food level		high food level	
	length of development	weight	length of development	weight
Pterostichus versicolor 19°C				
N	39	39	33	33
X	48.9	28.6	40.7	34.4
S	4.6	5.9	3.0	4.1
Calathus melanocephalus 15.5 °C				
N	15	15	8	8
X	187.7	11.8	142.4	16.3
S	6.0	1.3	15.4	1.2

Table 4. *C. melanocephalus.* The number of eggs laid by each female at different temperatures in 1976; * = females that died in winter during their stay within enclosures inside the field. A: Individual females ranked according to the numbers of eggs laid at different temperatures. B: The average numbers of eggs laid by females that died or survived the next winter.

A Temperature				♀♀ Died in winter
8.5°C	12°C	15.5°C	19°C	76/77
0				
	2			
7				
10 – – – – – –	– – – – – – – – –	– – – – – – – –	*	
	15			
	17			
21				
		29 – – – – – – – –	*	
36				
37 – – – – – –	– – – – – – – – –	– – – – – – – –	*	
38				
40				
45			45 – – –	*
	46			
		52		
	53 – – – – – –	– – – – – – – – –	*	
	55			
62				
	65			
		76 – – – – – – – –	*	
		80	80	
	81			
	82			
		84		
		93		
		93		
			99	
		105		
			109	
		110 – – – – – – – –	–	*
	115			
			125	
			132	
			134	
			143 – – –	*
		148 – – – – – – – –	–	
			151	
			180	

B The mean number of eggs of ♀♀ that died: 72.3±50.35 (in 1976).
The mean number of eggs of ♀♀ that survived: 72.4±45.98 (in 1976).

Mortality of all developmental stages of P. versicolor and C. melanocephalus
Extreme moisture conditions of the substrate (dry or wet peat-dust) were only unfavourable to the larvae and pupae of *P. versicolor* at low temperatures (12°c or lower) but at wide range of temperatures in *C. melanocephalus*. Total mortality during the whole development from egg until adult is much higher in *C. melanocephalus* than in *P. versicolor* under almost all conditions of temperature and substrate moisture, but most pronouncedly so under wet conditions (Fig. 4) (Van Dijk and Den Boer 1992). Hence the larvae of *C. melanocephalus* survive best under medium moisture conditions of the substrate combined with relatively low temperatures. *P. versicolor* on the other hand, survives best during development at relatively high temperatures and with medium or wet moisture conditions of the soil.

The inverse relationship between reproduction and survival
Different reproductive efforts in groups of females of *P. versicolor* and *C. melanocephalus* in the laboratory (induced by keeping them at different conditions did not result in differences in winter mortality when the beetles were released in the field at the end of the summer. There was no inverse relationship (trade-off) between reproductive output and survival in the field until the next breeding season (= the loss of individuals during winter) (Tables 1 and 4 A + B) (Van Dijk 1979a,b, 1994). Moreover Table 4A distinctly shows the positive relationship between temperature and size of egg production.

Following the same group of 30 females of *P. versicolor* during three succeeding years under the same conditions (during the breeding season in the laboratory at 19°c and ad libitum of food and subsequently – in autumn and winter – within enclosures in the field) it appeared that the older the beetles are (1, 2 or 3 years old), the more eggs the females are laying. Moreover the chance of death of a female in winter during these three years was not related to the previous reproduction size (Table 5).

Table 5. The number of eggs laid in three succeeeding years by *P. versicolor* females starting egg laying in 1976 and belonging to different survival categories.

survival category	1976	1977	1978
eggs laid by surviving females to next year	73.7 ± 61.4 (n=25)	97.1 ± 101.0 (n=20)	161.4 ± 85.0 (n=15)
eggs laid by females dying next winter	96.2 ± 79.0 (n=5)	110.4 ± 83.7 (n=5)	32.0 ± 32.5 (n=5)

Discussion

As shown temperature and soil moisture have an important influence on the size of egg production and the growth and survival of all developmental stages (Aukema 1991, Van Dijk 1992, 1994). Consequently the pattern of fluctuations in subpopulations of carabids is highly correlated with the weather (Baars and Van Dijk 1984, Van Dijk and Den Boer 1992). Yet a multiple regression analysis between weather factors and survival/recruitment rates shows that the explained variance in these rates decreases with the increase of the number of years (Baars and Van Dijk 1984). This paper shows that food is a significant factor for egg production and for growth and survival of all developmental stages. It has frequently been reported that the presence of food, particularly food of high quality, for carabid species in the field is limited both for larvae and adults (Ernsting et al. 1992, Heessen 1980, Juliano 1986, Lenski 1984, Nelemans 1987, 1988, Sota 1985, Szyszko 1990, Van Dijk 1979a, 1986a,b). The body weights and the size of egg production together are good estimators of the food conditions in the field (Van Dijk 1986a, Szyszko 1990). Given the relative shortage of food in the field, the size of egg production and the survival of the developmental stages highly depends on the quality of food (i.e. its nitrogen content). As shown in *B. tetracolum* small quantities of nitrogen-rich (i.e. more nutritious proteins) food (like eggs) consumed per female per day result in a much higher reproductive output than when high quantities of nitrogen-poor food were consumed. Moreover this paper shows that the quality of food also determined the maximum quantities of ingested food per female per day. It is often impossible to gather high quantities of nutritious proteins by eating nitrogen-poor food because of a restricted gut capacity (Mols 1993). Moreover, the low quantities of the maximum consumed nitrogen-rich food (1-2 mg/day) show that, beside the gut capacity the nitrogen content (i.e. the amounts of nutritious proteins) is of influence on the amounts of food ingested by carabids. Polyphagous carabid beetles eat many types of food (Hengeveld 1980) depending on the presence and distribution of their preys in the field. The quantity and the quality of these different kinds of prey will be variable and consequently the size of egg production will be variable in time and places. By switching between different kinds of food carabids enlarge the chance to get more high quality food (i.e. more nutritious proteins or more essential amino acids). Vermeulen and Szyszko (1992) and Wallin et al. (1992) showed indeed an increase in the size of egg production in carabids when they are able to switch from one prey to another. The concluded that "the more the amino acid composition of the food resembled that of the amino acid composition of the eggs produced the higher the egg production".

All these data confirm the hypothesis of White that animals usually experience a more or less continuous shortage of suitable food "specifically that which contains enough nitrogen" (White 1978, 1993). Food is the main limiting factor of animal numbers (Dempster and Pollard 1981).

No inverse relationship (trade-off) between reproductive output and survival in the field until the next breeding season (Van Dijk 1979a,b, 1994) for *P. versicolor*. The field ageing and survival of adults is determined by a composite of unpredictable environmental factors and physiological and physical conditions of the adults. Hence to postulate one simple inverse relationship between reproduction and survival as a compensatory mechanism leading to population, stability neglects the complexity of field situations and exaggerates the significance of such direct causal relationships. A possible advantage of a high egg production can be fully nullified again by a high larval mortality brought by unfavourable weather conditions. Nevertheless the fact in itself that females reproduce in more than one year enhance the stability of carabid populations living under varying environmental conditions (Den Boer 1968, Van Dijk 1982).

Acknowledgement

I am very grateful to Prof. Dr. L. Brussaard for critically reading the manuscript, Arnold Spee for invaluable assistance in all experiments and to Marian Hemmes-Stevens for typing the manuscript. Laboratory experiments on *Bembidion tetracolum* were carried out by Theo Cuijpers and René Verburg.

References

Aukema, B. 1991. Fecundity in relation to wing-morph of three closely related species of the *melanocephalus* group of the genus *Calathus* (Coleoptera: Carabidae). *Oecologia* 87: 118-26.

Baars, M.A. & van Dijk, Th.S. 1984. Population dynamics of two carabid beetles at a Dutch heathland. I. Subpopulation fluctuations in relation to weather and dispersal. *J. Anim. Ecol.* 53: 375-88.

Den Boer, P.J. 1968. Spreading of risk and the stabilization of animal numbers. *Acta Biotheor.* 18: 165-94.

Dempster, J.P. & Pollard, E. 1981. Fluctuations in resource availability and insect populations. *Oecologia* 50: 412-16.

Ernsting, G., Isaaks, J.A. & Berg, M.P. 1992. Life cycle and food availability indices in *Notiophilus biguttatus* (Coleoptera, Carabidae). *Ecol. Entom.* 17: 33-42.

Heessen, H.J.L. 1980. Egg production of *Pterostichus oblongopunctatus* (Fabricius) (Col. Carabidae) and *Philonthus decorus* (Gravenhorst) (Col., Staphylinidae). *Neth. J. Zool.* 30: 35-53.

Hengeveld, R. 1980. Polyphagy, oligophagy and food specialization in ground beetles (Coleoptera, Carabidae). *Neth. J. Zool.* 30: 564-84.

Juliano, S.A. 1986. Food limitation of reproduction and survival for populations of *Brachinus* (Coleoptera, Carabidae). *Ecology* 67: 1036-45.

Lenski, R.E. 1984. Food limitation and competition: a field experiment with two *Carabus* species. *J. Anim. Ecol.*: 203-16.

Mols, P.J.M., van Dijk, Th.S. & Jongema, Y. 1981. Two laboratory techniques to separate eggs of carabids from a substrate. *Pedobiol.* 21: 500-1.

Mols, P.J.M. 1993. Walking to survive; searching, feeding and egg production of the carabid beetle *Pterostichus coerulescens L. (= Poecilus versicolor Sturm). Agricultural University Wageningen Papers* 88 (3) and 93 (5): 1-203.

Murdoch, W.W. 1966a. Population stability and life history phenomena. *Amer. Nat.* 100: 5-11.

Murdoch, W.W. 1966b. Aspects of the population dynamics of some marsh Carabidae. *Journal of Animal Ecology* 35: 127-56.

Nelemans, M.N.E. 1987. Possibilities for flight in the carabid beetle *Nebria brevicollis (F.).* The importance of food during larval growth. *Oecologia* (Berlin) 72: 502-9.

Nelemans, M.N.E. 1988. Surface activity and growth of larvae of *Nebria brevicollis (F.)* (Coleoptera, Carabidae). *Neth. J. Zool.* 38 (1): 74-95.

Nelemans, M.N.E, den Boer, P.J. & Spee, A.J. 1989. Recruitment and summer diapause in the dynamics of a population of *Nebria brevicollis (F)* (Coleoptera, Carabidae). *Oikos* 56 (2): 157-69.

Price, P.W. 1984. *Insect Ecology.* John Wiley & Sons, New York.

Sota, T. 1985. Limitation of reproduction by feeding condition in a carabid beetle, *Carabus yaconicus. Res. Pop. Ecol.* 27: 171-84.

Stearns, S.C. 1992. *The evolution of life histories.* Oxford University Press, Oxford.

Szyszko, J. 1990. Planning of prophylaxis in threatened pine forest. *Biocenose based on an analysis of the fauna of epigeic Carabidae.* Warsaw Agricultural University Press, Warsaw.

Van Dijk, Th.S. 1979a. Reproduction of young and old females in two carabid beetles and the relationship between the number of eggs in the ovaries and the number of eggs laid. In: den Boer, P.J., Thiele, H.U. & Weber, F. (eds.) *On the Evolution of Behaviour in Carabid Beetles*, pp. 167-83. Miscellaneous Papers, Agricultural University Wageningen, 18.

Van Dijk, Th.S. 1979b. On the relationship between reproduction, age and survival in two carabid beetles: *Calathus melanocephalus L.* and *Pterostichus coerulescens L.* (Coleoptera, Carabidae). *Oecologia* (Berlin) 40: 63-80.

Van Dijk, Th.S. 1982. Individual variability and its significance for the survival of animal populations. In: Mossakowski, D. & Roth, G. (eds.) *Environmental Adaptation and Evolution*, Gustav Fischer, Stuttgart, 233-51.

Van Dijk, Th.S. 1986a. On the relationship between availability of food and fecundity in carabid beetles: How far is the number of eggs in the ovaries a measure of the quantities of food in the field? *Feeding Behaviour and Accessibility of Food for Carabid Beetles*, Warsaw Agricultural University Press, Warsaw, 105-21.

Van Dijk, Th.S. 1986b. How to estimate the level of food availability in field populations of carabid beetles. In: den Boer, P.J., Luff, M.L., Mossakowski, D. & Weber, F. (eds.) *Carabid beetles, their adaptations and dynamics*, Gustav Fischer, Stuttgart, 371-82.

Van Dijk, Th.S. & den Boer, P.J. 1992. The life histories and population dynamics of two carabid species on a Dutch heathland. I. Fecundity and the mortality of immature stages. *Oecologia* 90: 340-52.

Van Dijk, Th.S. 1994. On the relationship between food, reproduction and survival of two carabid beetles *Calathus melanocephalus* and *Pterostichus versicolor. Ecol. Entom.* 19 (3): 263-70.

Vermeulen, H.J.W. & Szyszko, J. 1992. Cooperation within a study of *Pterostichus oblongopunctatus F.* (Coleoptera, Carabidae). The influence of food and habitat quality on the egg production. *Proc. IVth European Congress*, Gödöllö, 1991. Hungarian Natural History Museum, Budapest: 592-601.

Wallin, H., Chiverton, P.A., Ekbom, B.S. & Borg, A. 1992. Diet, fecundity and egg size in some polyphagous predatory carabid beetles. *Ent. Exp. & Appl.* 65: 129-40.

White, T.C.R. 1978. The importance of a relative shortage of food in animal ecology. *Oecologia* (Berlin) 33: 71-86.

White, T.C.R. 1993. The inadequate environment. *Nitrogen and the abundance of animals.* Springer Verlag, Berlin.

Survival and reproduction in relation to habitat quality and food availability for *Pterostichus oblongopunctatus* F. (*Carabidae*, Col.)

J. Szyszko[1], *H.J.W. Vermeulen*[2] & *P.J. den Boer*[2]

[1]Department of Zoology Warsaw Agricultural University, 02-528, Rakowiecka 26, 30, Poland
[2]Biological Station Wijster, Kampsweg 27, 9418 PD Wijster, Drenthe, The Netherlands

Abstract
Observations were made of the Carabid fauna in more than a dozen stands in Holland, Poland and Germany. Special attention was paid to *P. oblongopunctatus*. Everywhere possible for this species was estimated: the male-to-female ratio; individual biomass at the moment of capture, after 24 hrs of starvation and after 24 hrs of feeding; respiration; consumption; number of eggs in ovaries; number of eggs laid; age structure of adults and the period of activity. Differences in the characteristics studied were found between the groups of individuals from different stands. Frequently there were differences between neighbouring stands, and these were more pronounced than those between stands some hundred kilometres apart. It is suggested that, with successional changes of habitat *P. oblongopunctatus* changes its life history pattern in a way that becomes apparent by larger body dimensions, lower respiration and consumption, a shorter lifespan of the adults, a larger number of eggs in the ovaries and different activity of males and females giving rise to differences in the sex ratio in the pitfall trap. The supposition is advanced that the main factor governing the way of life is food. An estimate of food conditions for larvae and adults respectively may be derived from the dimensions of the adult and the ratio of males to females in pitfall traps.

Key words: Biomass, habitat changes, consumption, egg production, food, respiration, *Pterostichus oblongopunctatus*, sex ratio, size

---

Arthropod natural enemies in arable land · II *Survival, reproduction and enhancement*
C.J.H. Booij & L.J.M.F. den Nijs (eds.). *Acta Jutlandica* vol. 71:2 1996, pp. 25-40
 ISBN 87 7288 672 2

Introduction

Survival and reproduction are two basic, interlinked phenomena which guarantee the existence of the species and influence the population dynamics. Much attention has been paid to these phenomena in scientific works and it has been demonstrated that survival and reproduction are modified by both abiotic, mainly temperature and moisture, and by biotic ones, mainly food, its accessibility and quality. With *Carabidae* this has been demonstrated both under laboratory and field conditions, and a review of this knowledge may be found in the works of Van Dijk & Den Boer (1992) and Van Dijk (1994).

In spite of the extensive information accumulated, it is difficult to establish which of these factors is decisive under natural conditions, and the main cause of this lack of simple but exact indicators allowing the phenomena in the field to be recorded.

The direct reason to the present study was the repeated demonstration that individuals of the same species of *Carabidae* from various forest habitats, differing in stand age and past use of the soil, are characterized by different dimensions (Szyszko 1978), varying biomass (Szyszko et al. 1978), different proportions of males and females in surface traps (Szyszko 1976, 1977) and different numbers of eggs in the ovaries (Borkowski & Szyszko 1984). On the assumption that the above mentioned characteristics might be indices of differences in survival and reproduction, it was decided to compare one of the most common of the forest *Carabidae* – *Pterostichus oblongopunctatus* in various forest stands in several European localities.

Materials and Methods

Collection of material

Material was collected in 16 forest stands (Szyszko et al. 1992), two of which were situated in Holland (near Wijster in 1986), five in north-west Poland (near Tuczno in 1987), and nine in southern Germany, with six near Bayreuth in 1988 and three in the Bavarian National Park in 1990. In each country the stands were visually examined and the only criterion for selection was the presence of *P. oblongopunctatus*. A brief description of the stands is given in Table 1. In all stands observations were made in the period of activity of the adults of *P. oblongopunctatus* by catching individuals (of the interaction group in the sense of Den Boer 1968, 1977) in live traps. The traps differed between countries on account of the different soils and different numbers of beetles. In Poland, a representative sample could be obtained from four 1-m catching grooves in the soil bordering a square with 10 m sides. In contrast the stony soils and small numbers of insects in Germany made it necessary to use 9-m plastic fences buried vertically into the soil with ten cylinder traps distributed along the sides. In all countries traps were inspected every three days. In the laboratory all the *Carabidae* individuals were counted according to species and their live biomass was determined. Further work was restricted to *P. oblongo-punctatus* and, depending on the availability of a laboratory or the possibility of killing beetles, concerned following features:

– the numbers of males and females

– the individual (mg) biomass of males and females at the moment of capture, after 24 hrs of starvation in the laboratory and then after 24 hrs of feeding in the laboratory with larvae of insects as food

– respiration of males and females in 24 hrs, estimated as the mg difference between biomass at the moment of capture and at the end of the period of starvation

– the consumption of males and females in 24 hrs, estimated as the difference between biomass after feeding and at the end of starvation

– the number of eggs deposited in the laboratory (in the starvation + feeding periods)

– the surface of the left elytron in mm^2 (determined by measuring length and breadth)

– the age of individuals, divided into young and old (from preceding years) specimens

– the number of eggs in ovaries

– the period of activity in days for males and females (the end of the period was considered to be the first day of the absence of individuals from traps after the period of culmination of numbers, with the restriction that this absence continued for three successive collection periods.

Determination of the age of individuals and of the numbers of eggs made it necessary to kill the beetles, so that attention could be paid to the degree of development of the flight muscles of the wings and to the presence of parasitic *Nematoda*.

A list of observations in the different stands (of the interaction groups) is given in Table 1. All field work and work on biomass estimation was done in Holland and Poland by H.J.W. Vermeulen, in Bayreuth (Germany) by J. Szyszko and in the Bavarian National Park by N. Schaffer from Bayreuth University. Elytra measurements, age determinations, establishment of the

Table 1. Number of males (M) and females (F) of *Pterostichus oblongopunctatus* caught and studied in the different stands for sex ratio, mean surface of elytra, biomass, consumption, respiration, number of eggs in the ovaries, number of eggs laid and age classes. N = Netherlands, PL = Poland, G = Germany.

Nr. of stand	Country	Description of stand	Individuals caught M+F	Sex-ratio		Mean surface of left elytra		Mean biomass		Mean consumption & respiration		Number of eggs in the ovaries	Number of eggs laid	Age class	
				M	F	M	F	M	F	M	F	F	F	M	F
1	NL	Oak-birch 30 years old	615			93	230					229		49	229
2	NL	Oak-pine 70 years old	250			88	198		198			75		87	75
3	PL	Pine 43 years old	752	294	458	35	35	71	70	71	70	101		99	101
4	PL	Pine 52 years old	599	249	350	32	46	52	105	52	105	110		68	110
5	PL	Pine 27 years old	434	177	257	55	70	63	90	63	90	100		77	100
6	PL	Pine 43 years old	326	114	212	46	64	58	81	58	81	88		70	88
7	PL	Pine 95 years old	89	37	52	12	25	17	28	17	28	18		21	18
8	G	Multispecies 4 years old	6	3	3						not analysed				
9	G	Spruce 90 years old	89	36	53	32	45	33	48	33	48	45	48	32	45
10	G	Beach-oak 140 years old	124	30	94	24	76	17	52	17	52	76	52	24	76
11	G	Oak 4 years old as undergrowth in Pine	52	19	33	18	17	17	23	17	23	20	23	19	20
12	G	Pine 35 years old	27	3	24	3	22	3	20	3	20	23	20	3	23
13	G	Pine 100 years old	51	16	35	15	29	15	32	15	32	29	32	15	29
14	G	Multispecies 2 years old	4	2	2						not analysed				
15	G	Natural mixed forest	29	15	14			15	14	15	14				
16	G	Windfall in natural forest	32	16	16			16	16	16	16				

numbers of eggs in ovaries and observations concerning the development of flight muscles and parasitism were made by H.J.W. Vermeulen.

Data collected

Because of the complicated processes occurring within each interaction group studied, the high degree of variability between individuals and the fact that the aim was to detect trends and regularities, it was decided to concentrate upon the data for *P. oblongopunctatus* obtained from each forest stand (one interaction group) for the whole period of activity of adults. Therefore we will restrict ourselves to indices such as sex ratio, the percentage of young individuals, and arithmetic means for the surface area of elytra, biomass, respiration, consumption, the number of eggs in ovaries etc., estimated for samples of individuals from different stands. Coefficients of correlations between the characteristics mentioned were established. Data from two stands in Germany were excluded from the study since as few as four and six individuals of *P. oblongopunctatus* were found there (Table 2). The authors thus had 12 samples for sex ratio, 13 samples for biomass at the moment of capture for females and 12 for males, 12 samples for biomass of males and females, 12 for separating young from old males and females, 12 for number of eggs in ovaries, 5 for eggs laid in the laboratory and 14 for determining the period of activity in days for males and females. These data are listed in Tables 1 and 2, together with the numbers of individuals from which they were obtained. A description of the carabid fauna in the stands (Table 2) was made on the basis of mean individual biomass (MIB), i.e. the quotient of the total biomass of all carabid individuals caught during the observation period and their numbers. According to Szyszko (1987, 1990), the higher the MIB value in milligrams, the higher the contribution of individuals of forest species and of species with large body dimensions that are characteristic for advanced stages of succession. Szyszko applies this characteristic as an indicator of the state of development of the fauna, with higher MIB indicating more advanced development of the fauna of *Carabidae* and a more advanced state of succession.

Results

Compilation of all data for the particular interaction groups of *P. oblongopunctatus* from each of the different stands shows a high degree of

Table 2. Short description of results obtained in the different stands from different countries. Full explanation in the text. * Data from the paper of Szyszko et all. 1992

Nr of stand	MIB *	Sex ratio M/F	Mean surface of left elytra in mm2		Mean biomass at moment of catching in mg		Mean biomass after 24h of starving in mg		Mean biomass after 24h of feeding in mg		Mean transpiration in mg		Mean consumption in mg		Mean number of eggs in the ovaries	Mean number of eggs laid	Time of activity in days		Percentage of young beetles		Number of beetles with flight muscles		Number of beetles with parasites	
			M	F	M	F	M	F	M	F	M	F	M	F	F	F	M	F	M	F	M	F	M	F
1	111		14.0	15.5											5.81		84	84	55.1	50.7				
2	126		14.1	15.6		54.2									4.85		70	56	58.6	38.7				
3	156	0.64	14.5	16.0	55.1	65.6	52.4	61.5	60.8	73.2	2.7	4.1	8.4	11.7	6.06		69	69	89.9	68.3				
4	193	0.71	14.5	16.0	52.8	65.1	50.8	62.3	62.3	72.9	2.0	2.8	11.5	10.6	6.37		69	69	79.4	59.1				
5	119	0.69	13.9	15.5	52.3	63.6	50.3	60.8	60.8	74.0	2.0	2.8	10.5	13.2	7.33		60	60	87.0	69.0				
6	269	0.54	14.6	16.4	53.1	63.6	51.4	60.7	60.2	74.4	1.7	2.9	8.8	13.7	7.44		60	60	74.3	58.0				
7	243	0.71	15.2	16.5	51.5	64.2	51.0	62.2	59.8	72.8	0.5	2.0	8.8	10.6	8.50		66	66	66.7	66.7				
8	63												not analysed											
9	215	0.68	15.5	17.3	62.4	73.9	59.2	69.6	64.9	78.2	3.2	4.3	5.7	8.6	8.82	0.44	23	23	75.0	64.4	4	3	1	
10	277	0.32	14.5	17.1	56.5	73.0	54.6	69.8	59.1	76.7	1.9	3.2	4.5	6.9	10.70	0.92	17	44	70.8	57.9			1	6
11	257	0.58	15.2	18.0	58.2	73.3	56.3	70.6	61.4	77.1	1.9	2.7	5.1	6.5	8.35	0.61	16	51	78.9	65.0		1		1
12	383	0.13	16.1	17.9	63.3	77.1	62.3	73.4	66.6	79.1	1.0	3.7	4.3	5.7	15.43	1.40	4	21	100	69.6		1	1	
13	310	0.46	16.1	18.3	64.7	75.4	62.6	72.5	67.1	79.6	2.1	2.9	4.5	7.1	7.86	0.62	24	43	86.7	72.4	1	1		
14	68												not analysed											
15	193	1.07			66.0	81.2	62.3	75.1	75.3	87.9	3.7	6.1	13.0	12.8			16	34						
16	95	1.00			62.7	77.6	59.0	69.9	71.3	88.8	3.7	7.7	12.3	18.9			46	48						

variability, a variability which is frequently higher among stands close together than among those in different countries. For instance, the sex ratio of males to females varied between 1.07 and 0.13, with both extreme values being recorded in German stands. The other interaction groups of *P. oblongopunctatus* from both Poland and Germany gave intermediate values (Table 2).

Mean values for the elytron surface area and for biomass at the moment of capture, after starvation and after feeding also showed great differences between the different stands. Mean biomass was in general higher for stands in Germany than for those in Poland but this was not observed for the surface area of elytra. In the latter case the mean values were similar for some stands in both countries especially for males (Table 2). The means obtained for the stands in Holland were within the range of variability of the means obtained in Polish stands. High differences between groups of individuals from various stands were also found in the estimates of respiration and consumption. It is difficult to say that the groups of individuals from Poland and Germany differed in consumption. Variability was very high and the highest and lowest consumptions were observed in groups of individuals from German stands (Table 2). The contribution of young individuals in the groups from different stands also varied. The lowest figure was for the stands in Holland. The range of variability was about similar in Polish and German stands.

Variability was also high with regard to the mean number of eggs in the ovaries, and also in this case the means from various countries frequently differed less than those from stands close together. Data for the number of eggs laid in the laboratory could only be obtained from five stands in Germany, but in this case differences were observed too. Means for the number of eggs laid per female within 48 hrs of starvation + feeding varied from 0.44 to 1.40 for different stands (Table 2).

As Table 2 shows, the activity of males and females was also found to differ among stands. It should be noted that, in spite of important differences in the period of activity between the different stands in Germany, the activity period was generally shortest there as opposed to that in the other countries. The differences in activity in stands in Poland were similar to those observed in Holland.

It was mentioned in the method section that attention was paid to the development of flight muscles and to parasitism by Nematoda. Developed flight muscles were observed in eleven young individuals from four of the German stands (Table 2). Nematoda were found in ten old individuals, also

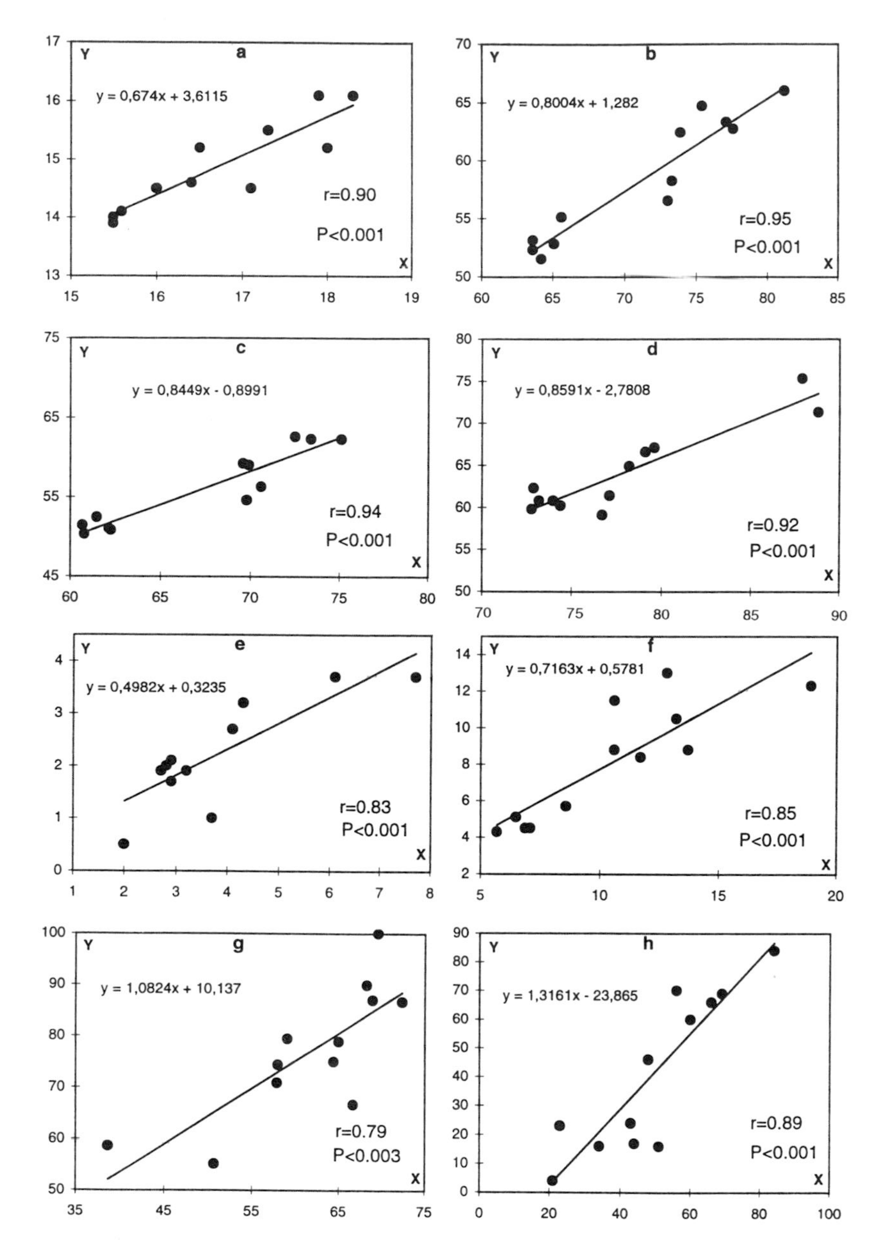

Fig. 1. Relationship between females (X) and males (Y) of *P. oblongopunctatus* in samples of individuals from various stands in Germany, Holland and Poland for surface area of elytra (a), biomass at the moment of capture (b), biomass after a period of starvation (c), biomass after feeding (d), respiration (e), consumption (f), the contribution of young individuals (g) and the period of activity (h).

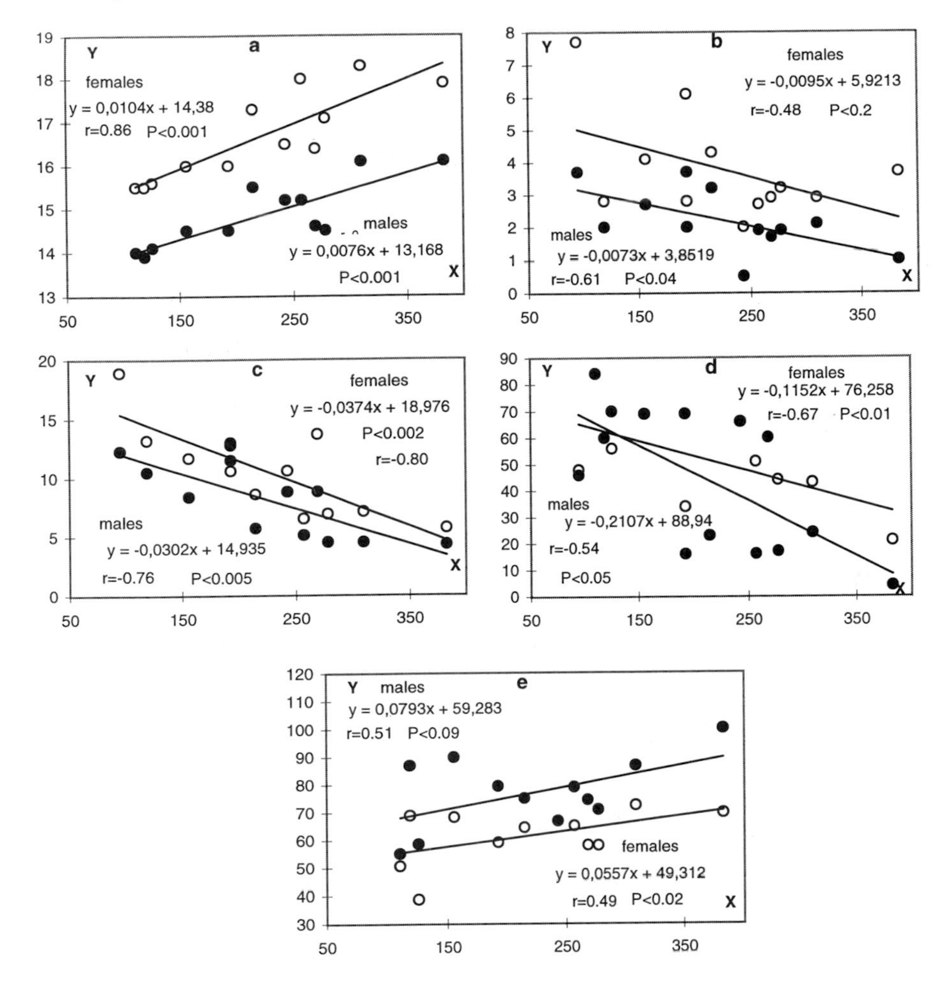

Fig. 2. Relationship between MIB (X) and the species characteristics (Y) obtained for groups of males (black circles) and females (open circles) of *P. oblongopunctatus* from various stands in Germany, Holland and Poland. a - surface area of elytra, b - respiration, c - consumption, d - period of activity in days, e - percentage contribution of young individuals.

from four stands in Germany. These results are interesting as *P. oblongopunctatus* had previously been considered to be usually unable to fly (Den Boer, 1987), while Nematoda parasitism of the species had not been mentioned sofar in ecological papers.

MIB values calculated for the different stands (Table 2) show that the interaction groups of *P. oblongopunctatus* examined were part of various faunas. Differences were high with values ranging from 63 to 383 mg, and with these extreme values again being recorded in German stands. The faunas of the other stands also showed large differences in MIB values, albeit within this interval. It must be noted, however, that it was difficult to catch a representative sample of individuals in the stands with the highest and lowest MIB values respectively.

The material above indicates that groups of *P. oblongopunctatus* differ from one another, and that the differences probably are not due to climatic conditions, since the characteristics of neighbouring stands frequently differed much more than those of widely-separated ones in different countries. Thus, other factors must have played a role and had similar effects on both males and females, since high correlation coefficients were found between females and males from the stands concerning size, biomass, respiration, consumption, the contribution of young individuals and the time of activity (Fig. 1). If this argument is correct, the causes of the observed differences should be in the habitat of the particular interaction groups, and to get a first indication correlation coefficients were calculated for the relationship between MIB and the size of elytra, consumption, sex ratio and the number of eggs in ovaries (Fig. 2). Lower correlation coefficients, (but usually above 0.5) were found in relation to respiration, the contribution of young individuals and the activity period. The correlations showed that the higher the MIB value, the larger the dimensions of the adults, the higher the number of eggs in the ovaries, the higher the proportion of young individuals, the lower the contribution of males in traps, the lower the respiration and consumption and the shorter the activity period.

Discussion

According to the suggestion of Szyszko (1987, 1990), each species would occur in a definite interval of successional development, as indicated by MIB values in the case of Carabids. This suggestion seems to be supported by the materials presented here, because *P. oblongopunctatus* was difficult to catch when MIB was low or high respectively (Tables 1,2). The correlation coefficients calculated for the relationship between MIB values and the studied characteristics in the different interaction groups from various stands seem to indicate that changes occurring in a habitat along with successional development are accompanied not only by changes in the abundance of species but also by changes in special characteristics of populations also.

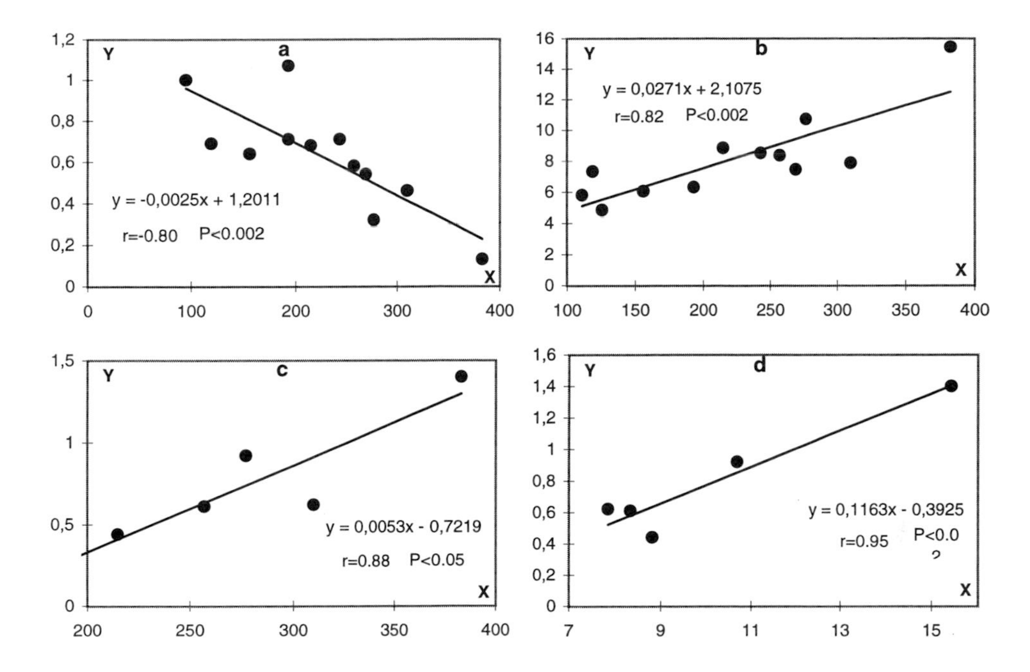

Fig. 3. Relationship between MIB (X) and particular characteristics estimated for samples pf *P. oblongopunctatus* individuals (Y) from various stands in Germany, Holland and Poland, a - sex ratio, b - number of eggs in ovaries, c - number of eggs laid in laboratory in relation to the numbers of eggs in the ovaries (X) and d - numbers of eggs laid in laboratory in periods of starvation and of feeding combined (Y).

The positive correlation found between MIB and the dimensions of elytra (Fig. 2) would seem to suggest that, with successional development, the adult of *P. oblongopunctatus* becomes larger and larger. Assuming that abiotic factors did not play a significant role in the data presented here, it may be supposed that it is the feeding situation for the larvae that plays the main role, since it has been demonstrated that better nutritional situations for larvae are associated with larger adults (Nelemans 1987a, Van Dijk 1984). If this is true then the size of the adult may be a simple indicator of the food situation for larvae in previous years. The positive correlation found between MIB and the percentage contribution of young individuals indicated that in the course of succession, the mean life period of the adults becomes shorter, or to put it in another way, the mortality of old individuals increases. It may be that some role here is played by parasites which have been detected in small numbers,

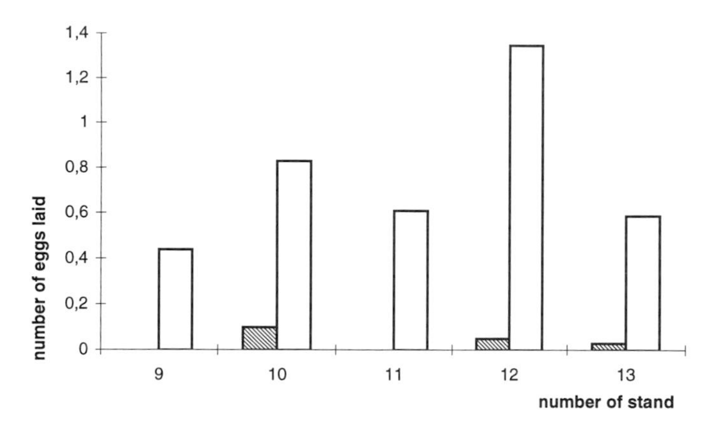

Fig. 4. Mean number of eggs laid in laboratory in periods of starvation (black bars) and feeding (open bars) by samples of females from various stands in Germany. Number indication of stands as in Table 1.

but especially in the stands where the MIB value was high. However,it was also shown, that the beetles were capable of flying giving the possibility that some individuals disperse by flying away from this habitat.

Assuming with Makowski & Szyszko (1986) that in the laboratory consumption by insects is inversely related to the degree of satiation of individuals, it may be said that the better the food situation in the habitat for the adult the more advanced the succession in the habitat. This is suggested by the negative correlation found between MIB and consumption by the imago (Fig. 2). A similar conclusion may be reached when considering the number of eggs in ovaries as an estimate of the food situation, since a positive correlation was found between MIB and the values for this estimate (Fig. 3). This was confirmed additionally, though only in relation to five stands, by a positive correlation between the number of eggs laid in laboratory conditions and the number of eggs in the ovaries (Fig. 3). Since the number of eggs laid may be considered a good indicator of the food situation, the presented data would seem to indicate unequivocally that the more advanced the developmental stage of the habitat in which the *P. oblongopunctatus* adult lives the better the situation for the adult.

However, this argument raises some doubts because of a negative correlation found between MIB and respiration (Fig. 2), (indicating that the higher the MIB value, the smaller the decrease in individual biomass in the starvation period). According to Makowski and Szyszko (1986) this would seem to be evidence of a deteriorating nutritional situation, since under constant laboratory conditions low respiration characterizes hungry individuals and high respiration those that are satiated (Makowski & Szyszko 1986). Also the fact that females with a large number of eggs in the ovaries, (originating from stands with a high MIB value) laid practically no eggs during the starvation period in laboratory conditions (Fig. 4) and only laid after food had been supplied (Fig. 4), might point in this direction. This

Table 3. Schematic representation of changes occuring in *Pterostichus oblongopunctatus* with changes in its habitat.

low state of development of fauna		high state of development of fauna
-high number of species	--->	-low number of species
-small species	--->	-big species
-small MIB	--->	-big MIB
	Interaction group of *Pterostichus oblongopunctatus*	
-long activity period	--->	-short activity period
-long survival of adult	--->	-short survival of adult
-complicated age structure	--->	-simple age structure
-small individuals (adult)	--->	-large individuals (adult)
-large proportion of males in pitfall traps	--->	-large proportion of females in pitfall traps
-low number of eggs in ovaries	--->	-high number of eggs in ovaries
-high number of eggs laid ?	--->	-low number of eggs laid ?
-good food conditions for adults ?	--->	-bad food conditions for adults ?
-bad food conditions for larva ?	--->	-good food situation for larva ?
-not able to fly ?	--->	-able to fly ?
-uneconomic way of life ?	--->	-economic way of life ?

Population of *Pterostichus oblongopunctatus*

-asynchronously fluctuating interaction groups	→	-synchronously fluctuating interaction groups	→	-asynchronously fluctuatig interaction groups
-small probability of high fluctuation of population size	→	-high probability of high fluctuation of population size	→	-small probability of high fluctuation of population size
-resistant population	→	-not resistant population	→	-resistant population

seems to indicate that the females from habitats advanced in succession have a large number of eggs in their ovaries, because they had accumulated these and a shortage of food in the field prevented them laying these.

If the latter argument is correct, then the negative correlation with MIB shown for the sex ratio in the traps (Fig. 3) can be a simple indicator of the food situation for the adult. The worse the food situation for the adult the greater the domination of females in the traps. this suggestion is confirmed by the paper of Szyszko & Gryuntal (in preparation), which for *Carabus hortensis* shows that well-nourished males are more active than females in the same situation, and, conversely, the hungry females are more active than hungry males. Such a situation will influence the ratio of males to females in traps.

The materials and suppositions presented here suggest that changes in the overall food situation for the larvae and adults of *P. oblongopunctatus* arise in the course of successional changes occurring in the habitat. The situation for larvae would improve with increasing MIB while for the adult it deteriorates. This seems to be confirmed by the fact that only young individuals possess functional wings in habitats with high MIB values, a fact that Nelemans (1987b) considered to indicate a good feeding situation for the larvae. the absence of flight muscles in old individuals from this these habitats may be the result of their disappearing with time because of a poor feeding situation for the adult.

In spite of the exact, but unknown, causes of the observed phenomena, it may be safely stated that successional changes in the habitat are accompanied by modifications in the life-history pattern of *P. oblongopunctatus*. In environments with more advanced stages of succession, lower respiration and consumption, shorter activity times, larger numbers of eggs in ovaries and probably smaller numbers of eggs laid, with larger dimensions of adults would all seem to be evidence of a much more economical way of life, which contrasts with that in habitats with less advanced stages of succession.

The discussion presented aand additional suggestions of Szyszko (1987, 1990), led to the scheme for changes in interaction groups and populations of *P. oblongopunctatus* in connection with the successional changes occurring in its habitat (Table 3). This scheme should be considered a working basis for further study which suggests, as announced in the paper by Den Boer et al. (1993) a necessity for genetic research in this species in habitats differing in their state of successional development. It can easily be imagined that *P. oblongopunctatus* may be present for no more than a dozen years in young forest plantations (after clear cut, Szyszko 1990) out of the thousands of years in natural stands.

Acknowledgements

The present work arose in collaboration with many people. We express many thanks to the Pila (Poland) and Bayreuth (Germany) Forest Administrations for the organization of the field experiments, to Prof. Thomas Bauer and Prof. Helmut Zwolfer for supplying funds for this research and to Dr. Werner Arends, Eng. Henryk Andrzejewski, Eng. Jan Krzyszkowski, Eng. Stanislaw Tomczyk, Wim Antoons, Thomas Bauman, Gunter Freese, Kathrin Ihen, Steffi Knoll, Anja Rot, Norbert Schaffer and Ulla Wigmann for help in the organisation work. We also wish to thank Krzysztof Klimaszewski for preparing the figures and tables.

References

Borkowski, K. & Szyszko, J. 1984. Number of eggs in the ovaries of some Carabidae (Col.) species in various pine stands on fresh coniferous forest habitats. *Ekol. Pol.* 32: 141-54.

Den Boer, P.J. 1968. Spreading of risk and stabilization of animal numbers. *Acta Biother.* 18: 165-94.

Den Boer, P.J. 1977. Dispersal power and survival. Carabids in a cultivated countryside. *Misc. Pap. 14 Landbouwhogeschool.*, Wageningen.

Den Boer, P.J., Szyszko, J. & Vermeulen, R. 1993. Spreading of extinction by genetic diversity in populations of the carabid beetle *Pterostichus oblongopunctatus* F. (Coleoptera, *Carabidae. Neth. J. Zool.* 43: 242-59.

Makowski, M. & Szyszko, J. 1986. On the possibility to use individual biomass to investigate the accessibility of food for *Carabidae* (Coleoptera). In: Den Boer, P.J., Grum, L. & Szyszko, J. (eds.) *Feeding behaviour and accessibility of food for carabid beetles.* Warsaw Agricultural Univ. Press, Warsaw: 123-30.

Nelemans, M.N.E. 1987a. On the life-history of the carabid beetles *Nebria brevicollis* (F.). Egg production and larval growth under experimental conditions. *Neth. J. Zool.* 37: 26-42.

Nelemans, M.N.E. 1987b. Possibilities for flight in the carabid beetle *Nebria brevicollis* (F.). the importance of food during larval growth. *Oecologia* 72: 502-9.

Szyszko, J. 1976. Male-to-female ratio in *Pterostichus oblongopunctatus* Col., *Carabidae)* as one characteristics of population. *Pedobiology* 16: 51-57.

Szyszko, J. 1977. Male-to-female ratio in *Carabus arcensis* (Col., *Carabidae*) in various forest environments. *Bull. Acad. Pol. Sc.* 25: 371-75.

Szyszko, J. 1978. Some remarks on the biometric characteristic of population of selected *Carabidae* (Col.) species in various forest habitat. *Pol. Pis. Ent.* 48: 49-65.

Szyszko, J. 1987. How can the fauna of *Carabidae* be protected in managed pine forest? *Acta Phytopath. entom. Hung.* 22: 293-303.

Szyszko, J. 1990. *Planning of prophylaxis in threatened pine forest biocenoses based on an analysis of the fauna of epigeic Carabidae.* Warsaw Agricultural Univ. Press, Warsaw.

Szyszko, J. & Gryuntal, S. (in prep.). Activity of *Carabus hortensis* (Col., *Carabidae*) in relation to forest habitat and food situation.

Szyszko, J., Szujecki, A., Mazur, S. & Perlinski, S. 1978. Seasonal changes in mean biomass of *Carabus arcensis* Hbst. and *Calathus erratus* (Sahlgb.) (Col., *Carabidae*) individuals in fresh forest pine stand. *Ekol. Pol.* 26: 297-307.

Szyszko, J., Vermeulen, H.J.W. & Schaffer, N. 1992. Cooperation within Europe in a study of *Pterostichus oblongopunctatus* F. (Coleoptera, *Carabidae*). I. The influence of habitat on the size. *Proceedings of the 4th ECE/XIII SIEEC*, Gödölö 1991: 584-91.

Vermeulen, H.J.W. & Szyszko, J. 1992. Cooperation within Europe in a study of *Pterostichus oblongopunctatus* F. (Coleoptera, *Carabidae*). The influence of food and habitat quality on the eggs production. *Proceedings of the 4th ECE/XIII SIEEC,* Gödölö 1991: 592-601.

Van Dijk, Th.S. 1994. On the relationship between food, reproduction and survival of two carabid beetles: *Calathus melanocephalus* and *Pterostichus versicolor. Ecol. Entom.* 19: 263-70.

Van Dijk, Th.S. & den Boer, P.J. 1992. The life histories and population dynamics of two carabid species on a Dutch heathland. I. Fecundity and the mortality of immature stages. *Oecologia* 90: 340-52.

Can pitfall trap catches inform us about survival and mortality factors? An analysis of field data for *Pterostichus cupreus* (Coleoptera: Carabidae) in relation to crop rotation and crop specific husbandry practices

L.J.M.F. den Nijs, C.J.H. Booij, R. Daamen,
C. A. M. Lock & J. Noorlander

Research Institute of Plant Protection, P.O.Box 9060,
6700 GW Wageningen, The Netherlands

Abstract

With the assumption that pitfall trap data for carabid beetle can be used to estimate relative differences in density between fields, long term trapping data were used to analyse survival and mortality patterns for *Pterostichus cupreus* in winter wheat, sugar beet and potato in various crop rotations. Indices were calculated for winter survival and for the recruitment per female over summer. In the crop sequence sugar beet – winter wheat activity-densities tend to increase due a high recruitment per female in sugar beet where larvae and young adults develop undisturbed and subsequent overwintering conditions in winter wheat are good. In the reverse order activity-densities due to harvesting wheat, leads to losses and the long stubble and fallow period before the next crop appears, creates unfavourable winter conditions. It is concluded that soil cultivation and the presence of crop cover are the main factors driving population changes.

The phenology of the beetles differs from one crop to another as a response to crops' specific favourable and unfavourable conditions. Starvation experiments with field-collected beetles showed that survival chances of heavier beetles are slightly better than that of light ones but depend on the beetles origin.

Key words: survival, carabidae, sugar beet, winter wheat, crop rotation, population dynamics, starvation, cop cover, tillage, overwintering, activity-density

---

Arthropod natural enemies in arable land · II *Survival, reproduction and enhancement*
C.J.H. Booij & L.J.M.F. den Nijs (eds.). *Acta Jutlandica* vol. 71:2 1996, pp. 41-55
 ISBN 87 7288 672 2

Introduction

Natural enemies are considered to be important in suppressing outbreaks of insect pests in arable fields. Enhancing natural occurring predators and parasitoids is a basic strategy in the development of integrated farming systems. Local fluctuations in densities of populations are the net result of reproduction, dispersal and mortality. For beneficial arthropods most information on population fluctuations is based on analysis of successive samples at a particular site over a certain period of time and do not provide information on the underlying reproduction, mortality and dispersal. To estimate reproduction, direct estimates on fecundity and oviposition in the field have been made for beneficial arthropods (Zangger et al. 1994, Baars & van Dijk 1984a, Wallin 1987, Sunderland & Topping, this volume). In *Bembidion guttula* net recruitment per female has been studied by Helenius (1995). Dispersal has mainly been quantified by studying individual movement of beetles in relation to field margins, barriers and (re)colonization after chemical treatments (Wallin 1985, Welling 1990, Duffield and Baker 1990).

Data on mortality of carabids under field conditions appear to be even more scarce. Grüm (1975) determined mortality patterns of carabid species inhabiting woodlands using pitfall traps and Riedel (1992) and Hokkanen (1993) determined winter mortality of carabids in field conditions. Laboratory studies revealed that abiotic factors can be crucial for survival (Pölking & Heimbach 1992, Heimbach 1995, this volume). It is difficult to estimate mortality or to do life table studies in arable fields and identify the most important mortality factors, because frequent assessments of densities are hardly feasible. Most attention has been given to the lethal effects of pesticides on beneficials both in the laboratory (Moosbeckhofer 1983, Heimbach 1988) and in the field (Dixon & McKinlay 1992, Purvis & Bannon 1992).

With regard to the use of pitfall traps, Baars (1979) stated that within one species "a satisfactory linear relationship is present between the mean densities in several habitats and in different years and the numbers of beetles trapped." As we have similar indications (den Nijs & Daamen unpublished data), we argue therefore that within one species, variation of pitfall trap catches between fields reflects variation in densities. This holds only when activity of the carabids is assumed not to be too variable between crops and fields and trapping is continuous over the whole activity period. Only under these assumptions crop related differences in mortality patterns may be derived from pitfall trap data.

Mortality may be caused by many factors including weather, food availability and cropping practices and will also depend on the characteristic of the population like the developmental stage, density and condition of the individuals.

Habitat quality often has an impact on growth and feeding conditions. Therefore, biomass and weight are important indicator for the condition of an organism and related to reproduction and survival. For instance a positive relationship was shown between weight and reproduction of carabids by Mols (1988), van Dijk (1994) and Zangger et al. (1994). Likewise it is assumed that heavier individuals possess better opportunities to survive during a period when food is not available (van Dijk, 1994).

In this paper we analyse to what extent it is possible to estimate survival and mortality of *P. cupreus* by comparing pitfall trap catches of fields with different crops and rotations in three years. The assumption that weight influences survival has been investigated by a laboratory experiment and is also described here. The chosen model species *P. cupreus* is a dominant species in many agricultural fields of central and northwestern Europe and occurs at variable densities in different crops (Booij & Noorlander 1991, Kegel 1994, Lys & Nentwig 1991, Wallin 1985, 1987, Zangger et al. 1994).

Material and Methods

Fields and trapping

In spring and autumn of 1992, 1993 and 1994 the carabid *P. cupreus* was caught in fields of sugar beet, winter wheat or a potato crop. Field size varied between 10 ha and 45 ha. Table 1 shows the crop rotation in the sampled fields from 1991 to 1994 and the type of pitfall trap that was used. Each year temperature and relative humidity was recorded in two contrasting fields (sugar beet and winter wheat) just above soil level by a LIDE microclimate recorder. Crop husbandry practices of each field were recorded. Ten traps, half filled with 4% formalin, were placed in the centre of each field at a distance of 10 m from each other. The traps were emptied weekly.

Live-trapped beetles were used for the survival experiment and for reproduction experiments described elsewhere (den Nijs et al., in prep). Live traps were only opened in daytime on days with temperature of 18°c or higher and checked every hour for presence of beetles over a maximum of 12 hours. The amount of live traps used depended on the crop on that field and were based on the expected number of beetles caught, based on earlier experience (Booij & Noorlander, 1991). In beet fields there were 200 traps, in wheat fields 100 and in potato 160.

Weight measurements

Beetles caught alive on the 20th April 1993 in the two sugar beet fields (M and L) and winter wheat fields (E and K), and the formalin caught beetles on 19th August and 2nd and 9th September 1993, and 12th, 19th and 26th April 1994 in field K (sugar beet) and L (winter wheat) were weighed on an analytical balance (Sartorius R200D, precision 0.1 mg). Weights of formalin caught beetles were used to estimate original fresh weights according to the experimentally derived formula *fresh weight = 1.19 * formalin weight* and when needed, weight was expressed as relative fresh weight, defined as fresh weight divided by the elytra area in mm^2 (den Nijs et al. 1996).

In spring 1992 only live traps were used (Table 1). To make these spring data from life traps comparable to the formalin trap data, the numbers of beetles caught in live traps were adjusted for the size of the trap and extrapolated from hourly captures to weekly captures. Parallel trapping in live and formalin traps in 1993 confirmed that this procedure was quite correct although some bias may be present. Changes in numbers and weight were analysed by ANOVA.

Survival estimates

When activity-densities changes differ between different crops and activity is assumed to be similar, differences should be due to either migration or mortality. During the winter only adults of *P. cupreus* are present and these remain more or less stationary, hence changes in number of adults are thus

Table 1. The experimental fields with crop rotation and used pitfall traps. L = life trap; F = formalin trap; S = spring; A = autumn). (y.m. = yellow mustard as autumn green manure crop)

	1991		1992	S	A	1993	S	A	1994	S
A	flax	-	spring wheat	L	F	spring barley	F			
B	winter wheat	-	sugar beet	L	F	corn	F			
D	winter wheat	-	potatoes	L	F	fallow	F			
E	potatoes	-	sugar beet	L	F	w-wheat/y.m.	L	F	potatoes	F
F	winter wheat	-	potatoes	L	F	fallow	F			
G	oil seed rape	-	winter wheat	L	F	peas	F			
K	winter wheat	-	spring barley			winter wheat	L	F	sugar beet	F
L	oil seed rape	-	winter wheat			sugar beet	L	F	winter wheat	F
M			potatoes			sugar beet	L	F	winter wheat	LF

caused by mortality. The numbers caught in the next spring were divided by the numbers caught in the previous autumn and were addressed as the "*winter index*". For the years 1992/1993 and 1993/1994 the "*winter index*" was assessed for all fields. Due to activity differences between autumn and spring this index gives a relative measure of survival.

The ratio between numbers caught in spring and autumn of the same year was also assessed for the years 1992 and 1993, here indicated as "*summer index*". When assuming migration to be negligible, this ratio may be considered to represent the *recruitment* per female, since it includes both fecundity and mortality of eggs and larvae. By estimating changes over the whole generation from spring to spring, an "*annual index*" of the population is measured including reproduction, mortality, immigration and emigration during all stages and provide an index of population performance. This "*annual index*" was determined over the years 1992/1993 and 1993/1994.

It should be realized that these indexes do not give absolute estimates of survival and mortality since catches are strongly affected by season and weather. They merely give an idea about the intercrops-differences of population development.

Starvation experiment.
The first beetles caught in spring 1993 (20th April) were weighed and kept individually without food in petri dishes (ϕ 9 cm) with two moist filter papers on the bottom at 15°c. Once a week the filter papers were refreshed and water was added. Every day the amount of dead beetles was counted. Every dead beetle was weighed again, to obtain a so called dead-weight, and stored at -20°c. The length of the survival period is used as a measure for the physical condition of the beetle at the start of the season. The relationship between weight and survival period was analysed by ANOVA.

Results

Fig. 1 shows the numbers of beetles caught in the ten formalin traps in fields with wheat (A and G), sugar beet (B and E) and potatoes (D and F) in 1992. In 1993 other crops were grown on these fields. The changes in numbers from autumn 1992 to summer 1993 clearly indicate the influence of the crop on the activity-density of the beetles. In winter wheat the activity-density declines rapidly after harvest. In the crops following winter wheat beetles were generally numerous and active in early spring. In the beet fields

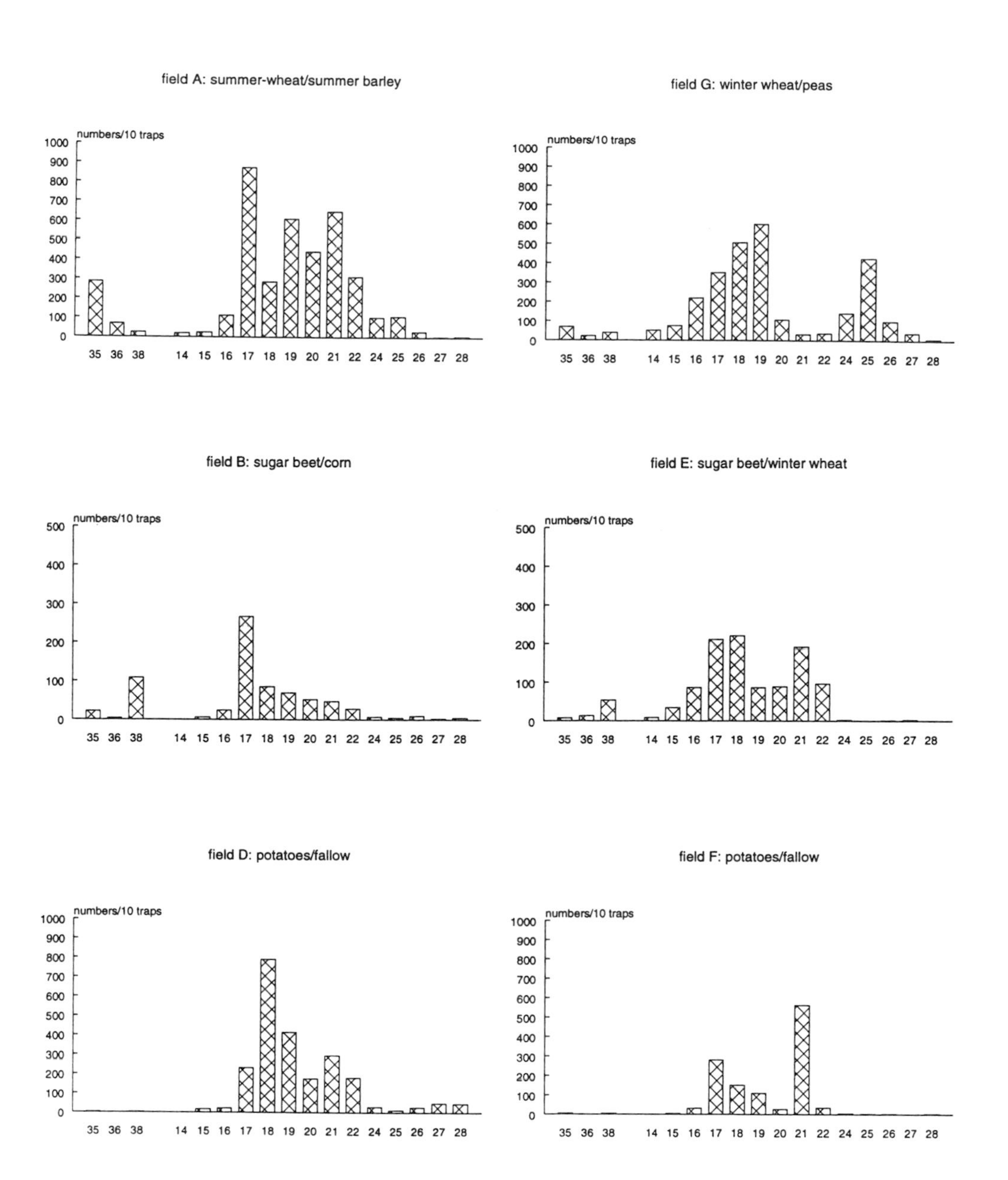

Fig. 1. Numbers of *Pterostichus cupreus* caught in formalin traps, expressed as total number per ten traps per week, in six fields with different crop rotations, during the autumn of 1992 and spring 1993.

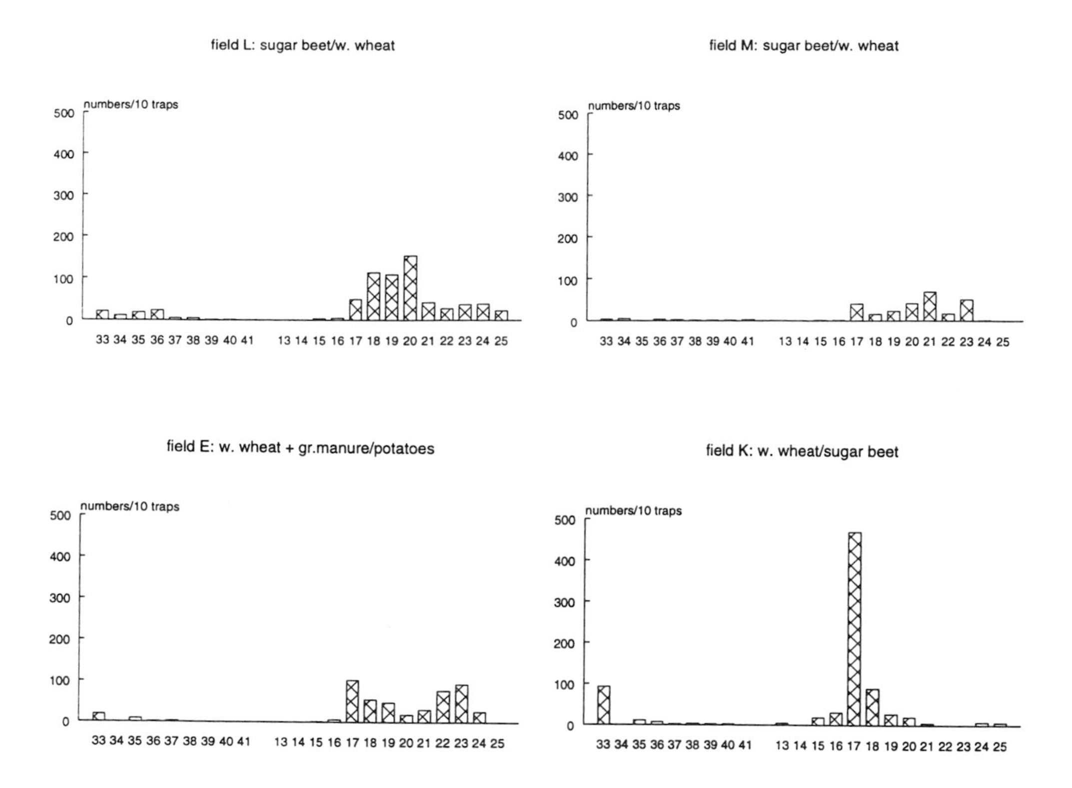

Fig. 2. Numbers of *Pterostichus cupreus* caught in formalin traps, expressed as total number per ten traps per week, in four fields with different crop rotations during the autumn of 1993 and spring of 1994.

captures are still rising in the autumn and in the following spring the population recovery is somewhat slower than after wheat. In the two potatoes fields hardly any beetles were caught during the autumn and here also activity densities increased much slower in spring next year compared to the fields with wheat as a precrop (A and G, Fig. 1). Thus it appears that the phenology of the beetles is influenced by the crop type and its associated management practices.

The numbers of beetles caught in autumn 1993 and spring 1994 in fields with sugar beet (L and M) and winter wheat (E and K, Fig. 2) were much lower than in 1992-93. This is presumably due to the very wet autumn and winter of 1993-94. Additionally, though average spring temperature was normal, the amount of sun radiation was very low and only a few warm days occurred during the autumn of 1992 and spring of 1993.(e.g. 26th April). This certainly had a negative influence on activity levels in all fields and thus on catch. Nevertheless, the same tendency was found in 1992-93. An "early crop" in the previous year is followed by an early appearance of the carabids in the next spring whereas a "late crop" like sugar beet results into a late appearance in the next spring.

Table 2 shows the annual index, the summer index and the winter index over the time periods spring 1992/1993 and 1993/1994. The general level of annual indices is lower in 1993/1994 than in 1992/1993. This may be mainly due to weather factors (see above). The influence of temperature on activity of the beetles is very clear on day 29 in April 1993 and day 26 in April 1994. These days were very sunny and warm resulting in high catches in all fields independent on crop type (Figs. 1 and 2).

Differences in annual indices between fields are clearly present, the lowest values are found in fields with winter wheat as pre-crop in both years whereas the highest values were reached in fields with sugar beet or potatoes as pre-crop. During the summer sugar beet and potatoes cover the soil with a full green crop whereas winter wheat is ripening, dying off and harvested in

Table 2. The annual index, summer index and winter index of *Pterostichus cupreus* in various fields with different crop rotations over the years.
annual index = spring population / previous spring population
summer index = autumn population / previous spring population
winter index = spring population / previous autumn population.

		annual index	summer index	winter index
	1992/1993			
A:	s-wheat/barley	1.5	0.15	9.46
B:	s-beet/corn	7.1	1.56	4.55
D:	potato/fallow	29.0	0.08	382.2
E:	s-beet/w-wheat	9.8	0.69	14.0
F:	potato/fallow	5.8	0.05	121.8
G:	w-wheat/peas	0.7	0.04	21.2
	1993/1994			
E:	w-wheat*/potato	0.4	0.03	15.23
K:	w-wheat/s-beet	0.2	0.03	5.60
L:	s-beet/w-wheat	0.9	0.12	7.25
M:	s-beet/w-wheat	0.6	0.04	14.28

* in this field green manure was grown after the harvest of the winter wheat.

the same period. This coincides with the most vulnerable period of the life cycle of the carabid, the larvae and pupae stage. The values of the summer index shows this more clearly: summer index is highest in fields with sugar beet in both years. Field M is an exception, in spring (29/4/1993) this field was sprayed with parathion against the pygmy-beetle (*Atomaria*), which had also a drastic effect on the *P. cupreus* population.

Annual index is also influenced by the survival of the carabids during the winter. Crop cover during winter and no tillage in spring should give rise to higher winter index-values. The results of 1992/1993 are in agreement with this. The high winter index-values in the potato-fallow crop rotation (field D and F), seem to be a result of the absence of soil cultivation in the fallow fields and very low numbers of caught beetles in the potato fields in autumn. The results of 1993/1994 show that green manure (field E) during the winter has the same effect as winter wheat (field L and M), namely crop cover during winter enhances the winter survival.

Table 3 shows the mean weight of beetles caught in the different fields at different times of the year. As expected, weight increases during the summer with the tendency (not significant, $P > 0.05$) that beetles reach higher weights in sugar beet than in winter wheat. During winter 1993-1994 a significant decrease in weight appears in bare field K, whereas the beetles in field L, in which winter wheat was sown, kept the same weight. It is likely that crop cover during winter affected the condition of the beetle and this influenced their survival positively, which is also reflected in the winter index (Table 2). Results of the starvation experiment showed that survival can be affected by weight.

Table 3. The mean weight with standard error between brackets, expressed in mg, of *Pterostichus cupreus* during the seasons 1993 and 1994 in four fields with different crop rotations. Field E: w-wheat/green manure followed by potatoes in 1994, field K: w-wheat followed by sugar beet, field L and M: sugar beet followed by w-wheat. Within the column different letters indicate significant differences at P=0.05 (ANOVA).

	1993 spring	1993 autumn	1994 spring
E	65.46(1.22) a		
K	66.67(1.05) a	77.88(2.74) *	69.33(2.20) a
L	66.92(0.85) a	81.32(2.33) ns	80.81(2.52) b
M	72.99(0.91) b		

Fig. 3a shows the mean relative weight of the beetles caught in fields with sugar beet or winter wheat in spring 1993. based on 64, 115, 135 and 115 beetles from fields E, K, (winter wheat) and L and M (sugar beet).The mean survival period and the mean weight loss due to the starvation are shown in Fig. 3b and 3c. The hypothesis that heavier beetles can survive longer in periods without food is clearly supported by these data. The beetles from field L stayed alive longer than was expected on basis of their mean relative weight. ANOVA on the survival period revealed that gender and field were highly significant (P<0.001). The weight loss during starvation is in accordance with the relative weight of the beetles; ANOVA on the dead-weight of the beetles showed no significant influence of the fields but the gender was highly significant (P<0.001), the mean relative dead-weight of the male of *P. cupreus* was 3.2 mg/mm^2 and for the female 2.9 mg/mm^2.

Discussion

Differences in numbers of beetles caught in the same field during the year are the result of the density of the beetles (which in itself is the sum of reproduction, migration and mortality), the physiological state of these organisms (dormancy, hibernation) and activity. Temperature is one of the driving forces for activity (Baars & van Dijk 1984b; Honèk 1988). Since different species have different temperature preferences, this makes comparisons between species very tricky. However, within one species differences in catches are likely to reflect differences in densities as long as there are no major other factors which cause activity differences. This assumption, which was used to analyse the data in this paper, is disputable and at least an oversimplification. However, even when the assumption is not fully valid, conclusions can be drawn under proper reservation.

In this research project we used this approach to analyse captures of *P. cupreus* in different fields, varying in size between 10 ha and 45 ha and with various crops. Another necessary assumption was that migration does not play an important role for *P. cupreus* in these large fields. There is no evidence that substantial flight occurs (Kegel 1994), nor that the beetles frequently move between fields which are separated by ditches as in our study. Therefore differences in captures are assumed to be caused by differences in reproduction and mortality or survival between the crops. From field-based reproduction studies (Den Nijs et al, in prep.) we draw the conclusion that between crops differences in the reproduction of *P. cupreus*

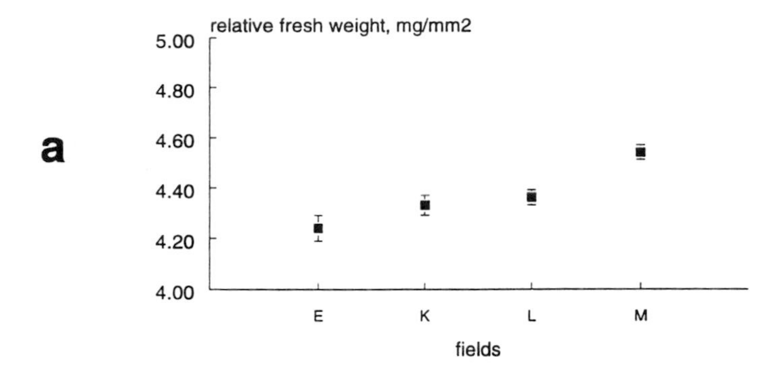

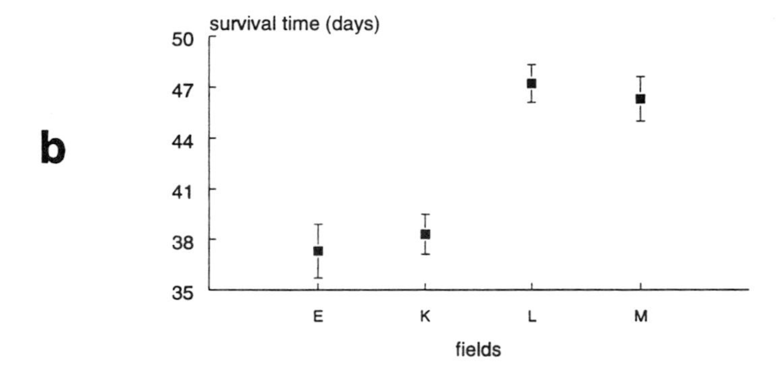

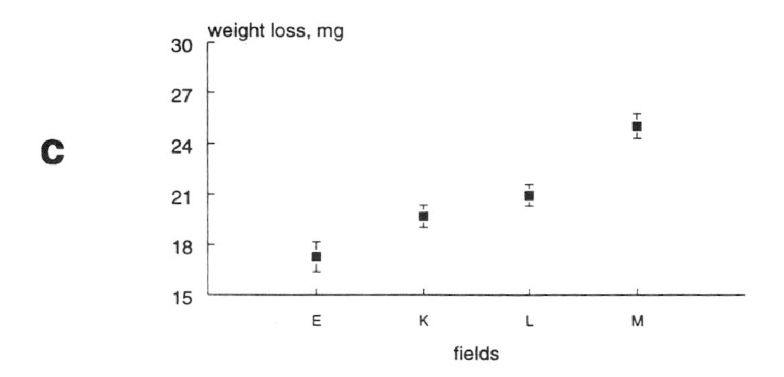

Fig. 3. The mean relative weight (3A), the mean survival period (3B) and the mean weight loss due to starvation (3C) of *Pterostichus cupreus* caught in winter wheat fields (E and K) and sugar beet fields (L and M) on the 10th April 1993. Precrops were sugar beet, spring barley, winter wheat and potatoes respectively.

were small. When variability in reproduction between fields is small and migration is negligible, differences in captures can be ascribed to mortality. Between the different fields we indeed found enormous differences in numbers of caught beetles (Fig. 1 and 2). Hence we conclude that these differences in activity-density are caused mainly by mortality.

There are many reasons why beetles die under field conditions. Entomophagous fungi and nematodes can play a role according to Hokkanen (1993) and Riedel & Steenberg (1992), although it seems to be generally low. Knowledge on mortality caused by predation or starvation in the field is scarce (Brunsting et al. 1986, Luff 1994). Mortality caused by abiotic factors like temperature, moisture or soil disturbance might be more important. We can only show indirect evidence for this. For instance, summer survival in all beet fields was high and thus mortality was low (Table 2) due to the crop cover. Luff (1994) stated after his research on starvation on carabid larvae that summer larvae are more likely to die early because of high temperature than winter larvae. Furthermore, he hypothesized that summer larvae may need to be in moist situations in order to survive. The measured temperatures in the wheat and sugar beet field differed only slightly in the period that larvae are present. The mean humidity in both crops at soil level reached 90% in June and July. After harvest of the wheat however, the humidity in the wheat field drops drastically. Since sugar beet is harvested much later humidity stays high until late in the autumn. This might have influenced the survival of the late instars. The suggestion of Luff (1994) that the abiotic factors temperature and humidity of the soil, which are affected by the crop type on the field, are main factors of mortality are slightly supported by our results.

Tillage is expected to have a high impact on the carabid population (Blumberg & Crossley, 1983). Comparative research on farming systems showed that the amount of carabid beetles increases when the farming system is less intensive (Booij 1992, Basedow 1991), although Büchs (1991) found that some species increase when intensity of soil cultivation increases. Our results show that soil disturbance by means of earthing up potatoes and the harvest of an early crop influences the summer index negatively (Table 2).

Crop cover during winter can provide shelter and food for the carabids that are active when temperature is sufficiently high. It is expected that survival of these carabids will be better in fields which have a cover crop than in bare fields. When carabids are in a good condition, which means that they are relatively heavy, this should increase their chances of survival (Ernsting et al., 1990). The circumstances before the winter determines the condition of the carabid. In an agroecosystem the presence of a crop or green manure in the autumn is therefore important. Hokkanen (1993) found that in determining winter survival of *Meligethes aeneus* in a field in Finland, the body weight in autumn was the main factor for survival and van Dijk (1994) found that the successful overwintering was related to the presence of good food in the field. We determined the body weight of the beetles after the winter, beetles that had survived the winter period, and these beetles were subjected to starvation. The positive relationship between body weight and length of survival period was evident, which corresponds with Hokkanen's findings.

In conclusion one may say that there are lots of pitfalls in estimating mortality from pitfall trap data. There is a strong need for methods which give a direct and absolute estimate of mortality in relation to particular factors. When this is not feasible it should at least be quantified how large and variable differences in activity in different crops can be to be able to estimate density changes and density differences between crops by pitfall trapping.

Acknowledgements

We thank D. Monsma and his assistants and G. Grimme for allowing us to work on their fields.

References

Baars, M.A. 1979. Catches in pitfall traps in relation to mean densities of carabid beetles. *Oecologia* 41: 25-46.

Baars, M.A. & Van Dijk, Th.S. 1984b. Population dynamics of two carabid beetles at a dutch heathland. I. Subpopulation fluctuations in relation to weather and dispersal. *Journal of Animal Ecology* 53: 375-88.

Baars, M.A. & Van Dijk, Th.S. 1984a. Population dynamics of two carabid beetles at a dutch heathland. II. Egg production and survival in relation to density. *J. Anim. Ecol.* 53: 389-400.

Basedow, Th. 1991. Population density and biomass of epigeal predatory arthropods, natural enemies of insect pests, in winter wheat fields of areas with extremely different intensity of agricultural production. *J. Plant Diseases and Protect.* 98: 371-77.

Blumberg, A.Y. & Crossley, D.A. 1983. Comparison of soil surface arthropod populations in conventional tillage, no-tillage and old field systems. *Agro-ecosystems* 8: 247-53.

Booij, C.J.H. 1992. Farming systems and insect predators. *Agric. Ecosysystems Environment* 40: 125-35.

Booij, C.J.H. & Noorlander, J. 1991. The impact of integrated farming on carabid beetles. *Proc. Exp. & appl. Entomol.* 2: 16-21.

Brunsting, A.M.H., Siepel, H. & Schaick-Zillesen, P.G. van 1986. The role of larvae in the population ecology of carabids. In: den Boer, P.J., Luff, M.L., Mossakowski, D. & Weber, F. (eds.) *Carabid beetles, their adaptation and dynamics.* Gustav Fischer, Stuttgart-New York, 399-411.

Büchs, W. 1991. Einfluss verschiedener landwirtschaftlicher Produktionsintensitäten auf die Abundanz von Arthropoden in Zuckerrübenfeldern. *Verh. Gesellschaft für Ökologie* 20: 1-12.

Dijk, Th.S. van 1994. On the relationship between food, reproduction and survival of two carabid beetles: *Calathus melanocephalus* and *Pterostichus versicolor. Ecol. Entomol.* 19: 263-70.

Dixon, P.L. & McKinlay, R.G. 1992. Pitfall trap catches of and aphid predation by *Pterostichus melanarius* and *P. madidus* in insecticide treated and untreated potatoes. *Entomol. exp. appl.* 64: 63-72.

Duffield, S.J. & Baker, S.E. 1990. Spatial and temporal effects of dimethoate use on populations of carabidae and their prey in winter wheat. In: Stork N.E. (ed.) *The role of ground beetles in ecological and environmental studies.* Intercept, Andover, Hampshire. 95-104.

Grüm, L. 1975. Mortality patterns in carabid populations. *Ekol. Polska* 23: 649-65.

Helenius, J. 1995. Use of emergence traps for teneral Carabidae in estimating recruitment rates in agricultural habitats. *Acta Jutlandica* LXX/2: 101-12

Heimbach, U. 1988. Nebenwirkungen einiger Fungizide auf Insekten. *Nachr. Deutsch. Pflanzenschutzd.* 40: 180-83.

Hokkanen, H.M.T. 1993. Overwintering survival and spring emergence in *Meligethes aeneus*: effect of body weight, crowding, and soil treatment with *Beauveria bassiana. Entomol. exp. appl.* 67: 241-46.

Honék, A. 1988. The effect of crop density and microclimate on pitfall trap catches of Carabidae, Staphylinidae (Coleoptera), and Lycosidae (Aranea) in cereal fields. *Pedobiologia* 32: 233-42.

Kegel, B. 1991 The biology of four sympatric *Poecilus* species. In: Desender et al. (eds.) 1994 *Carabid beetles: Ecology and Evolution.* Kluwer Academic Publishers. 157-63.

Luff, M.L. 1994. Starvation capacities of some carabid larvae. In: Desender, K., Dufrene, M., Loreau, M., Luff, M.L., & Maelfait, J-P. (eds.) *Carabid beetles, ecology and evolution.* Series Entomologica 51: 171-75. Kluwer Academic Publishers.

Lys, J.A. & Nentwig W. 1991 Surface activity of carabid beetles inhabiting cereal fields. Seasonal phenology and the influence of farming operations on five abundant species. *Pedobiologia* 35: 129-38.

Mols, P.J.M. 1988. Simulation of hunger, feeding and egg production in the carabid beetle *Pterostichus coerulescens L.* Agricultural University Wageningen papers 88-3 (1988).

Moosbeckhofer, R. von 1983. Laboruntersuchungen über den Einfluss einiger Pflanzenschutzmittel auf Ei- und Larvenstadien von *Poecilus cupreus L.* und *P. sericeus* Fischer d. W. (Col., Carabidae). *Z. ang. Ent. 95*: 513-23.

Nijs, L.J.M.F. den, Lock, C.A.M., Noorlander, J. & Booij, C.J.H. 1996. Search for quality parameters to estimate the condition of *Pterostichus cupreus* (Col., Carabidae) in view of population dynamic modelling. *J. Appl. Ent.* 120: 147-51.

Purvis, G. & Bannon, J.W. 1992. Non-target effects of repeated methiocarb slug pellet application on carabid beetle (Coleoptera: Carabidae) activity in winter-sown cereals. *Ann. appl. Biol.* 121: 401-22.

Riedel, W. 1992. Hibernation and spring dispersal of polyphagous predators in arable land. Ph. D. thesis, University of Aarhus, Aarhus, Denmark.

Wallin, H. 1985. Spatial and temporal distribution of some abundant carabid beetles (Coleoptera: Carabidae) in cereal fields and adjacent habitats. *Pedobiologia* 28: 19-34.

Wallin, H. 1987. Life cycles, reproduction and longevity of the medium-sized carabid species *Pterostichus cupreus L.*, *P. melanarius Illiger*, *P. niger Schaller* and *Harpalus rufipes De Geer* (Coleoptera: Carabidae) inhabiting cereal fields. In: Distribution, movement and reproduction of carabid beetles (Coleoptera: Carabidae) inhabiting cereal fields. Ph. D. thesis of Swedish University of Agricultural Sciences. Uppsala, Sweden.

Welling, M. 1990. Förderung von Nutzinsekten, insbesondere Carabidae, durch Feldraine und herbizidfreie Ackerränder und Auswirkung auf den Blattlausbefall im Winterweizen. Ph. D. thesis, University of Mainz, Mainz, Germany.

Zangger , A., Lys, J.A. & Nentwig, W. 1994. Increasing the availability of food and the reproduction of *Poecilus cupreus* in a cereal field by strip management. *Entomol. exp. appl.* 71: 111-20

Estimating the mortality rate of eggs and first free-living instar *Lepthyphantes tenuis* (Araneae: Linyphiidae) from measurements of reproduction and development

C.J. Topping[1] *& K.D. Sunderland*[2]

[1]Present address: Danish Environmental Institute, Kalø, Grenaavej 12, DK-8410 Rønde, Denmark (Farmland Ecology Unit, Department of Land Resources, S.A.C., Mill of Craibstone, Bucksburn, Aberdeen AB2 9TS, UK)
[2]Horticulture Research International, Worthing Road, Littlehampton, West Sussex BN17 6LP, UK

Abstract
Lepthyphantes tenuis is a highly successful species within the agroecosystem and as such is of significance as a polyphagous predator within these habitats. However, little is known about its population dynamics, and in particular mortality estimates have been elusive. This paper describes a method developed to determine the level of mortality suffered by the egg and first free-living instar by projecting the expected population density of early instar spiders and comparing this with the actual density found over the period of one season. Projected densities were calculated using observations on the fecundity of *L. tenuis* and by using simple development rate models based on temperature. Mortality estimates were obtained by comparing the predictions of a simple model of mortality for eggsacs and spiderlings to actual field density by least squares fit. Mortality of egg and first free-living instar was predicted to be approximately 93-96% during the period of this study.

Key words: development rate models, agroecosystem, polyphagous predator, fecundity

Introduction

Lepthyphantes tenuis is extremely successful at exploiting the ephemeral habitats found in the agroecosystem. This species is often the dominant spider

Arthropod natural enemies in arable land · II *Survival, reproduction and enhancement*
C.J.H. Booij & L.J.M.F. den Nijs (eds.). *Acta Jutlandica* vol. 71:2 1996, pp. 57-68
 ISBN 87 7288 672 2

in cereal crops (e.g. Sunderland 1991). It occurs in a wide range of habitats and has a very large geographical range, including most of Europe and New Zealand, where it is also dominant in agricultural habitats (Topping, unpubl.). In addition this species builds a large web and can take a large proportion of aphids in cereal crops (Fraser 1982, Sunderland et al 1986, Alderweireldt 1994). These factors make *L. tenuis* one of the most agriculturally important spiders.

This species is highly fecund and dispersive and as such demonstrates the characteristics of r-selected species typical of these habitats. As a consequence eggsac and hatchling mortality can be expected to be high, and will undoubtedly play a large part in the population dynamics of this species. There is a wide range of potential mortality factors both for the eggsacs and the hatchling spiders. These include eggsac parasitoids (e.g. *Aclastus* sp.) or egg parasitoids (e.g. *Baeus* spp.). Both of these species attack the eggs within the eggsac and effectively kill the whole unit (van Baarlen et al. 1994). In laboratory experiments we observed that mould resulting from too damp an environment often caused eggsac mortality, too dry an environment would be an equally efficient killer. It is also probable that larger predators, such as Carabidae, also take a proportion of eggsacs of this species, although this has not been recorded. Hatchling spiders will also be taken by predators (Sunderland et al. 1994), including other spiders, they will also be easily killed by extremes of temperature or moisture deprivation and probably also suffer mortality from rainfall due to their small size (<1 mm). As a result estimation of the size of the various mortality factors is extremely difficult.

This paper describes a method developed to determine the level of mortality suffered by the egg and first free-living instar by projecting the expected population density of early instar spiders and comparing this with the actual density found over the period of one season in a field of winter wheat in southern England during 1991.

Methods

The overall approach was a simple one. There are four factors which determine the number of hatchling spiders in the field at any point in time, these are births, deaths, dispersal and the speed of development. We have a measure of between-habitat dispersal for recently hatched *L. tenuis* and have estimated the overall effect of immigration to the field and emigration from

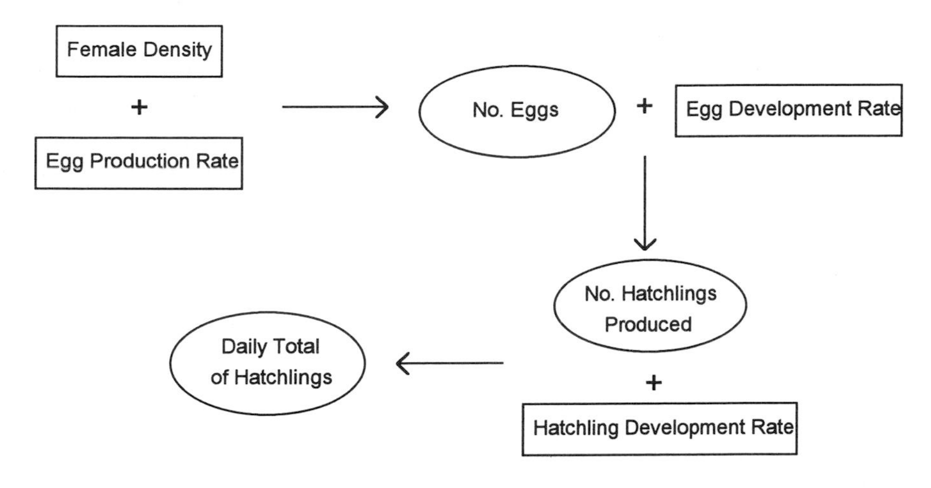

Fig. 1. Flow diagram illustrating the steps involved in calculating egg and hatchling mortality for *L. tenuis* in winter wheat during 1992.

the field to be approximately 10.95 spiderlings m^{-2} emigrating during the period of this study (Topping & Sunderland, in prep.) Hence if we can successfully measure births and developmental rates we can use this data to predict field density in the absence of mortality. The difference between actual and predicted field densities will provide an estimate of mortality plus emigration. Since we know overall emigration rates we can determine the overall level of mortality experienced by eggs and hatchlings in the population during this study (Fig 1).

Density estimates of adult female and hatchlings were taken from a population of spiders in a field of winter wheat in southern England during 1991. Estimates were obtained by approximately fortnightly sampling using D-Vac suction sampling and post searching by hand as described by Sunderland & Topping (1995), Topping & Sunderland (1994b). Temperatures were recorded continuously in the field by Squirrel datalogger using a probe inserted under a small clod of earth. Descriptions of methods of collecting data on egg-production and development of eggs and hatchlings are available elsewhere (Sunderland & Topping 1992, Sunderland et al. 1995) and are only briefly described below.

Eggsac production rate

Estimates of egg production were made by removing adult *L. tenuis* from the field and keeping them, unfed, for one week in Petri-dishes lined with moist filter paper. These were placed in the study field, under a Stevenson's screen, to protect them from direct sunlight. After this period the proportion of females producing eggsacs was determined. This procedure was repeated approximately weekly throughout the season. The total number of eggsacs produced was calculated by: (1) estimating the density of females present in the field by interpolation between the fortnightly density estimates; (2) calculating thc proportion of spiders producing eggsacs on each day by the same method; (3) multiplying the density of female spiders present on each day by the proportion expected to be producing eggs on that day to give the number of eggsacs produced per day.

Egg development rate

Eggsacs produced by incarcerated females were returned to the Stevenson's screen and monitored until they hatched. The length of time between eggsac production and hatching and the number of hatchlings produced was recorded. Temperatures within the Petri-dishes were recorded using a data-logger. It was found that the development rate was largely temperature-dependent. The relationship can be described by the model :

1) $$D = \frac{K}{T - t}$$ (Wigglesworth, 1950),

where D is the development time in days, K the thermal constant (number of day degrees required for development), t the threshold temperature below which no development takes place and T the actual temperature experienced.

The development rate was calculated as the invers of the developmental period (days^{-1}). Following the method used by De Keer & Maelfait (1987 & 1988), the mean temperature was plotted against development rate and a linear regression fitted. The intercept of this line provided 't', and 'K' was obtained by arithmetic manipulation. Using data from this study and others (Sunderland & Topping, in prep.) t was calculated to be 4.6°c with K at 244.5 day°c.

Using this equation and hourly temperature measures from the field, the development time for eggsacs produced on each day of the season was calculated. The date on which each eggsac would hatch could then be projected. The mean number of hatchlings produced from each eggsac was found to be constant at 21.28 during the experimental period, hence the number of hatchlings produced in the field on each day could be calculated.

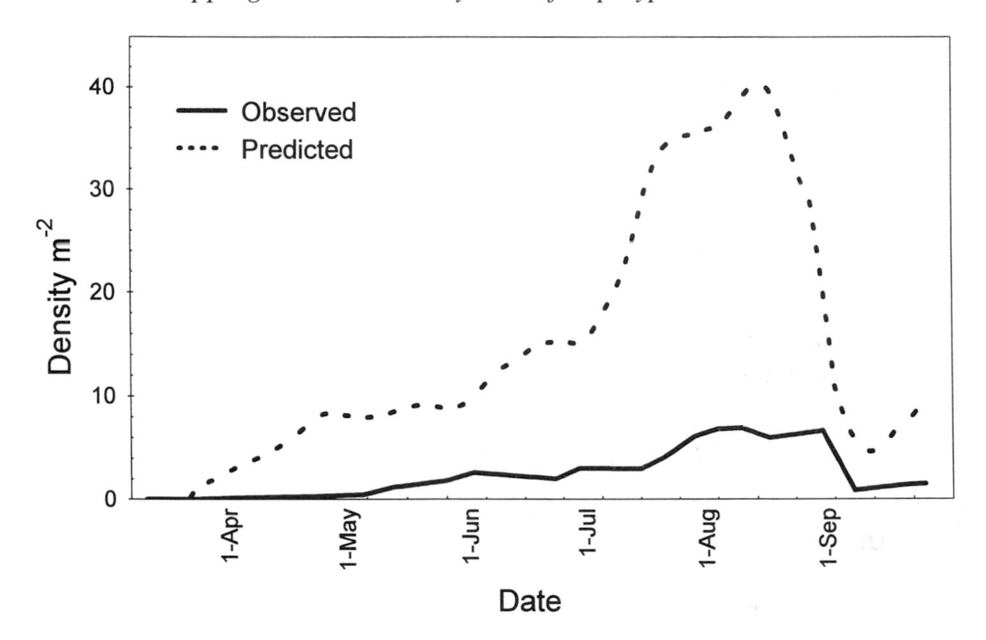

Fig. 2. Predicted and observed hatchling density in the absence of mortality and emigration for L. tenuis in a field of winter wheat during 1992.

Hatchling development rate

In order to be able to predict the number of hatchlings present at a particular date, the length of time hatchlings are present in the field must be calculated. This requires a knowledge of their development rate. De Keer & Maelfait (1987 & 1988) found that the development rate of immature *Erigone atra* and *Oedothorax fuscus* were under the direct influence of temperature following equation (1) above. Following their methodology temperature-dependent development rates were observed under three feeding regimes: flour mites *Acarus siro* L. (Acari), *Folsomia candida* (Willem) (Collembola: Isotomidae), *Folsomia candida* with pollen and yeast) and produced overall mean values of t & K for hatchling development (t=0.37°c, K=389.8 day°c) (Sunderland et al., in prep.). The development time for each daily batch of hatchlings was calculated from field temperature data and a predicted running total of hatchlings, which would be present in the field in the absence of mortality, was determined (Fig. 2).

Model construction

The model was constructed as a simulation following the steps described above (Fig. 1). Density data and reproduction data were available from 11th March 1991 until ploughing on 24th September 1991. Harvest occurred on 19th August and coincided with a 90% drop in hatchling numbers. In order to simulate the effects of harvest, a 90% mortality was therefore applied to any eggsacs or hatchlings present on that date. Due to low temperatures in the period following harvest a number of eggsacs would not hatch until the following year. There would also be a small number of eggsacs produced and hatched before measurements began in March. In order to compensate for these, eggsacs which survived up to the end of the year were put back into the population at the beginning of the year, before the calculation of hatchling density.

Incorporation of spider loss factors

Spider loss was incorporated for both eggsacs and hatchlings as a fixed daily percentage for each stage. Thus, the longer a batch of eggsacs or hatchlings took to develop, the smaller the overall survivorship for that batch.

Determination of best fit to actual density

This was achieved by:

1) Smoothing the absolute density measures using a three point running average
2) Obtaining the actual daily density of hatchlings by interpolation
3) Smoothing the predicted hatchling curve using a seven point running average
4) Transforming all data by $\log_{10}(n+1)$
5) Creating a squared difference estimate of fit by squaring the differences of the daily values
6) Repeating the analysis with the addition of a constant field temperature in order to take account of any difference in temperatures experienced by eggsacs and hatchlings and the field temperature probe
7) Determination of the best fitting variants of mortality by running simulations of a range of all possible values of eggsac and hatchling mortality and determining the least squares estimates for each (e.g. Fig 3).

Since later data (Sunderland, unpubl.) suggested that the maximum rate of development of hatchling *L. tenuis* could be twice that of the values used here, this analysis was repeated using a value of K_2 of 194.9.

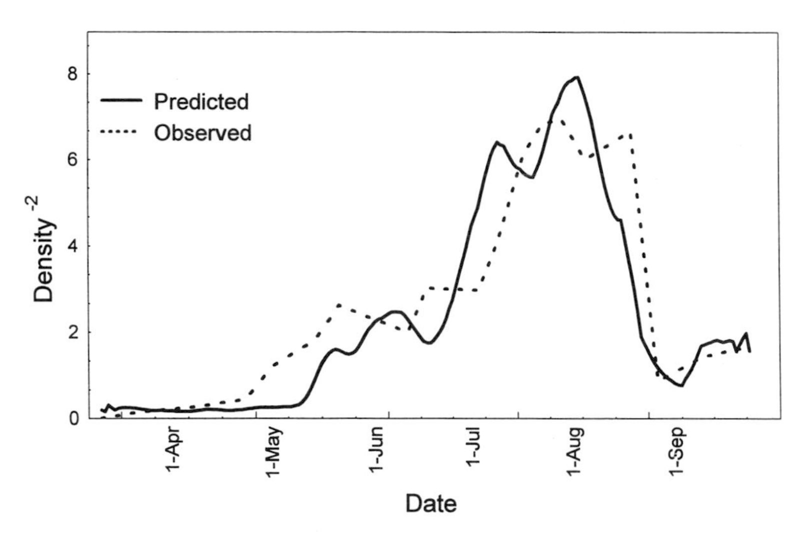

Fig. 3. Predicted and observed hatchling density after the estimation of egg mortality and hatchling loss for *L. tenuis* in a field of winter wheat during 1992.

Results

Sensitivity was investigated by assessing the deviation from the fit to the actual density curve when using basic starting parameters. These were t & K as calculated for hatchling and eggsac development with M_1 (Eggsac loss) and M_2 (hatchling loss) set at 0.006 and 0.21 respectively (best fit values). Each value was altered by 10 and 20% either side of the starting point (Table 1). When considering the fit to the actual density curve, the model was relatively insensitive to changes in the threshold temperature (t_2) and thermal constant (K_2) for the hatchlings. Most variation was obtained by altering t_1 & K_1 for eggsacs or by altering hatchling loss (M_2). However, if the loss factors are altered by +/- 0.01 or 0.02 rather than by +/- 10% and 20% of their initial values, greater sensitivity is shown to M_1 (Table 2).
Sensitivity was also tested by determining the total output of second instar spiders per unit area.Variations in K_2 and M_2 caused greatest fluctuations in the numbers of hatchlings surviving per unit area. The difference in spiderling production between +20% and -20% of M_2 was only 6.89 spiderlings m^{-2}, out of 394 eggs produced. At first sight varying t_1 and K_1 produced an unusual result with an increase in spiderling production for a 10% change in either direction, followed by a fall in production for a 20% change. However, this is explained by the fact that altering the timing of

hatchling production results in a greater or lesser number of hatchlings being present in slightly warmer periods. This would lead to faster development and lower losses.

Varying the constant added to the measured temperature suggests that the temperature recorded by the probe may have been 3°c lower than suggested by the minimum least squares fit (Table 3). It was assumed that this value represents the difference in temperature between the temperature probe, under a clod of earth, and the eggsacs & spiderlings in the lower portions of the vegetation.

For the best-fit scenario of M_1=0.006, M_2=0.21 and t=+3°c, the total spiderling deficit was 390.2 out of 394 eggs produced. Of these, 10.95 can be accounted for by emigration from the field, hence the total number killed was 379.25 m^{-2}. This is a mortality figure of 95.7% and represents a net loss from the population of 97.2%. Taking the maximum and minimum hatchling survival from Table 1 suggests that for a 40% variation in the most sensitive parameter the resulting mortality estimates were 94.5-96.3%. When using the doubled development rate estimate (K_2=194.9) the estimates of M_1 and M_2

Table 1. Least squares fit of model prediction to number of hatchlings measured after varying model parameters by 10% and 20% either side of best fit values.

	t_1	K_1	M_1	t_2	K_2	M_2
Initial Value	4.6	244.5	0.006	0.37	389.8	0.21
Fit to data						
Plus 20%	9.22	10.78	8.22	8.16	8.09	11.97
Plus 10%	8.70	9.60	8.16	8.16	8.11	9.41
No Addition	8.16	8.16	8.16	8.16	8.16	8.16
Minus 10 %	8.13	7.91	8.24	8.16	8.22	9.17
Minus 20%	8.35	8.42	8.41	8.16	8.26	12.90
Number hatchling produced						
Plus 20%	5.81	5.60	5.68	5.75	4.52	3.64
Plus 10%	6.36	6.63	5.74	5.76	5.02	4.52
No Addition	5.80	5.80	5.80	5.80	5.80	5.80
Minus 10 %	6.32	6.22	5.94	5.82	7.01	7.69
Minus 20%	5.94	6.10	5.92	5.84	8.74	10.53

were very similar (0.007 and 0.19 respectively; least squares fit 8.61). However, the estimate for total mortality was reduced to 93%, with a total hatchling production of 26.67 out of 394 eggs produced. These mortality figures represent the expected mortality based on the mean number of hatchlings produced per eggsac during 1991. There will therefore be a small additional loss due to individual egg or hatchling mortality before hatchling emergence.

Table 2. Least squares fit of model prediction to number of hatchlings measured after varying model parameters by 0.01 and 0.02 either side of best fit values.

	M_1	M_2
Initial Value	0.006	0.21
Fit to data		
Plus 0.02	20.97	9.16
Plus 0.01	12.52	8.45
No Addition	8.16	8.16
Minus 0.006	15.87	8.21
Number hatchling produced		
Plus 0.02	4.10	4.55
Plus 0.01	4.88	5.11
No Addition	5.80	5.80
Minus 0.006	6.42	7.59

Table 3. Estimates of daily egg mortality ($M_{1)}$, daily hatchling loss ($M_{2)}$ after varying temperature constants in the model.

Temperature Constant Added ºC	M_1	M_2	Squared Difference
-1	0.001	0.22	11.38
0	0.002	0.22	11.25
1	0.004	0.21	10.34
2	0.005	0.21	9.13
3	0.006	0.21	8.16
4	0.009	0.21	8.47
5	0.008	0.21	8.29

Discussion

The assumptions made in the construction of this model are varied and numerous. This reflects the difficulty of obtaining estimates of mortality which can have so many causes. The most obvious lack of realism concerns the fact that mortality is considered by the model to be constant throughout the season. This is certainly unlikely to be the case for any single mortality factor, especially parasitoid attack and weather, but in the absence of any better estimates a constant overall mortality factor is probably the most robust estimate. Further assumptions concerning the application of mortality equally over the duration of the instar are also unlikely to reflect the true situation. In the eggsac, attack by parasitoids is likely to occur early on after the eggsac is produced to give the parasitoid time to develop before the spiders hatch. Despite these uncertainties the resulting fit of the model to the actual density data is generally good with very little variation in the numbers of hatchlings produced for quite large variation in any individual parameter's value. Even when applying considerable variation in the two mortality parameters or doubling hatchling development rate, mortality estimates for eggs and first instar hachlings are upwards of 93%. At this rate of mortality each female spider would need to produce almost 200 eggs in order for the population to replace itself. Since observations suggest that a female can produce up to ten eggsacs (Sunderland et al. 1995) this would not be unreasonable, assuming much lower rates of mortality in later growth stages. It also suggests that the life-history strategy of *L. tenuis* is designed to maximise production of young. This coupled with high levels of adult dispersal provide a typically r-selected strategy which probably explains the success of this species in the ephemeral habitats of the agroecosystem.

Acknowledgements

This work was funded by SOAFD, MAFF, NERC and the Leverhulme Trust.

References

Alderweireldt, M. 1994. Prey selection and prey capture strategies of linyphiid spiders in high-input agricultural fields. *Bull. Brit. arachnol. Soc.* 9: 300-8.

Van Baarlen, P, Sunderland, K.D. & Topping, C.J. 1994. Eggsac parasitism of money spiders (Araneae: Linyphiidae) in cereals, with a simple method for estimating percentage parasitism of *Erigone* spp. eggsacs by Hymenoptera. J. *Appl. Ent.*: (in press).

De Keer, R. & Maelfait, J-P. 1987. Laboratory observations on the development and reproduction of *Oedothorax fuscus* (Blackwall, 1934) (Araneida, Linyphiidae) under different conditions of temperature and food supply *Rev. Biol. du Sol.*: 24, 63-73.

De Keer, R. & Maelfait, J-P. 1988. Laboratory observations on the development and reproduction of *Erigone atra* Blackwall, 1833 (Araneae, Linyphiidae). *Bull. Brit. arachnol. Soc.* 7: 237-42.

Fraser, A.M. 1982. The role of spiders in determining cereal aphid numbers. Ph.D. thesis, University of East Anglia.

Sunderland, K.D., Fraser, A.M. & Dixon, A.F.G. 1986. Field and laboratory studies on money spiders (Linyphiidae) as predators of cereal aphids. J. *Appl. Ecol.*: 23, 433-47.

Sunderland, K.D. 1991. The ecology of spiders in cereals. *Proc. 6th Int. Symp. Pests and Diseases of small grain cereals and maize.* Halle/Saale, Germany 1: 269-80.

Sunderland, K.D. & Topping, C.J. 1992. The spatial dynamics of spiders in farmland. *XII International Congress of Arachnology.* Mem. Queensl. Mus. 33: 639-44.

Sunderland, K.D., Ellis, S.J., Weiss, A., Topping C.J. & Long, S.J. 1994. The effects of polyphagous predators on spiders and mites in cereal fields. *Proc. BCPC Conf. - Pest & Diseases*: 1151-56.

Sunderland, K.D. & Topping, C.J. 1995. Estimating population densities of spiders in cereals. Proceedings of the 1st EC Workshop. Estimating population densities and dispersal rates of beneficial predators and parasitoids in agroecosystems. Aarhus 1993. *Acta Jutlandica* 70: 13-22.

Sunderland K.D., Topping, C.J. , Ellis, S., Long, S.J., Van de Laak, S. & Else, M. 1995. Reproduction and survival of linyphiid spiders, with special reference to *Lepthyphantes tenuis* (Blackwell). Proceedings of the 2nd EU Workshop. Wageningen 1994. *Acta Jutlandica*: (in press).

Topping, C.J. & Sunderland, K.D. 1994a. Methods for quantifying spider density and migration in cereal crops. *Bull. Brit. arachnol. Soc.* 9: 209-13.

Topping, C.J. & Sunderland K.D. 1994b. A spatial population dynamics model for *Lepthyphantes tenuis* (Araneae: Linyphiidae) with some simulations of the spatial and temporal effects of farming operations and land-use. *Agr. Eco. & Env.* 48: 203-17.

Wigglesworth, V.B. 1950. *Principles of Insect Physiology.* Methuen and Co., London.

Assessment of survival and mortality factors in field populations of beneficial arthropods

C.J.H. Booij[1], *C. Topping*[2], *J. Szysko*[3], *Th. van Dijk*[4], *M. Paoletti*[5], *J. Helenius*[6]

1 Research Institute for Plant Protection,
P.O. Box 9060, 6700 GW Wageningen, The Netherlands
2 National Environmental Research Institute,
Grenåvej 12 DK-8410 Kaløe, Denmark
3 Institute for Forest Protection and Ecology,
Rakowiecka 26-30, Warszawa PL-02528, Poland
4 Biologisch Station Wijster
Kampsweg 27, 9418 PD Wijster, The Netherlands
5 Dept. Biology, Padova University,
Via Trieste 75, 35121 Padova, Italy
6 Agriculture Research Centre, Institute of Plant Protection
FIN-31600 Jokioinen, Finland

Abstract
Measurement of survival and mortality and identification of factors involved is crucial for understanding population dynamics of beneficials insects in agro-ecosystems. Methodology of and bottlenecks in the study of mortality and survival were discussed as part of an EU workshop on enhancement and dynamics of beneficial arthropods. Practical considerations and approaches are summarized and the relative importance of different survival factors is discussed.

Key words: survival, mortality, beneficial arthropods, food, methodology habitat management, agroecology

Introduction

Theoretically, density changes in closed populations can be fully described by the parameters reproduction and survival. In open populations the additional parameter migration is needed. One of the basic problems in population dynamics is that all three parameters are variable in time and space and are strongly affected by local conditions and hence hard to predict.

Arthropod natural enemies in arable land · II *Survival, reproduction and enhancement*
C.J.H. Booij & L.J.M.F. den Nijs (eds.). *Acta Jutlandica* vol. 71:2 1996, pp. 69-78
 ISBN 87 7288 672 2

In highly mobile species the dynamics of local populations may be fully determined by fluctuations at a larger spatial scale and local reproduction and survival factors may relatively unimportant. Dynamics of different sub-populations are not necessary correlated (den Boer 1968), especially when dispersal is low and local conditions differ. Therefore the scale at which populations are studied and spatial interrelations between local populations are crucial aspects in population ecology. When studying the impact of spatial and temporal variation in survival and underlying mortality factors, these scale effects cannot be neglected. The recognition that many populations consist of interacting local populations may help to understand the mutual linkage between large and small scale dynamics and the factors involved at different scales (Topping & Sunderland 1994).

In the context of enhancement of beneficials there is not a prior argument to study reproduction and survival separately. Knowing the key factors which determine net reproduction from generation to generation (i.e. reproduction * survival), one may be able to manipulate conditions in order to increase population levels in the long run. Also in long-term population studies, which are almost lacking in agriculture, but see Jones (1979) and Luff (1982), this approach may be sensible.

In general however, it is felt that in the dynamic agricultural system a separate understanding of reproduction and survival factors is important to manipulate short term changes in order to achieve maximal natural control when speaking about beneficials. Furthermore, the timing of agricultural practices like pesticide applications, tillage and harvest in relation to the phenology of beneficials may affect the impact on beneficials enormously when they occur at a vulnerable phenological stage.

Obviously there is no single best approach for studying survival. A better understanding can only be achieved by studies at different levels of integration. These may range from large scale field studies, which correlate changes in numbers with multiple factors, to laboratory studies which determine the impact of a single factor on an individual or even a genotype. Models may be helpful in linking these different approaches (Topping and Sunderland 1994).

Assessment of survival by tracking changes in numbers

Life table data or time series of density estimates from the field (e.g. Wallin 1987) may be a firm base to understand major effects of agricultural practices on population dynamics. When collected in long-term studies, these data may be used for key-factor analysis though such an analysis for beneficials in agroecosystems is lacking. In experimental and comparative approaches covering one or a few seasons, similar data can be collected to estimate short term effects of particular factors.

In both cases some basic conditions have to be fulfilled in order to prevent unacceptable bias in estimating survival rates:
a) sampling methods should give a more or less unbiased estimate of absolute or relative density. In this context the measurement of activity densities may be misleading and should be interpreted with great care (Spence & Niemelä 1994).
b) emigration and immigration during the sampling period should be negligible or it should be possible to measure and quantify these processes. For many beneficial arthropods in agriculture, which are relatively mobile, study areas or experimental units should be several hectares or more to exclude substantial migration effects. In practise this constraint is often violated.
c) care should be taken that the whole activity period of relevant stages is covered by the sampling method. This is especially true when critical and sensitive moments in the life cycle have to be tracked or where sampling relies on activity which is strongly weather dependent.

It should be realized that most populations of beneficials in agriculture have to be considered as relatively open (ballooning spiders, syrphid flies, coccinellids, many carabids). The dynamic nature of agricultural system leads to major problems in accurately monitoring numbers throughout the season, in particular during critical moments like time of harvest and soil cultivation. Therefore the opportunities for detailed life table studies under field conditions are limited. Another problem may be to estimate numbers of eggs and juveniles which are often more difficult to sample (carabids) or hard to identify (spiders and staphylinids).

Despite all these limitations monitoring numbers under different agricultural regimes may provide us with many rough indications for critical survival factors (den Nijs et al. 1996). In ecotoxicology monitoring numbers before and after application is used as a standard technique to assess field toxicity (Jepson 1989a).

Experimental approaches to measure survival

Starting with simplified hypotheses generated from field work, more profitable and efficient approaches may be used to get a more detailed insight in survival. Since a multitude of factors involved may act on survival under field conditions, a critical selection is necessary.

From a practical point of view one may select, for example, only those factors which can be manipulated by the farmer. Taking into account all the biological information which is available for a particular species, one may be able to select the most critical and sensitive stages in the life cycle. Since each developmental stage is affected by a different and limited set of factors in the field in a restricted period, factorial experiments may be set up to quantify the impact of these factors. Such experiments may be performed in the laboratory or at (semi) field level. For instance effects of soil moisture on larval survival can be studied experimentally in lab and field conditions (Heimbach, this issue). If such a relationship is important and larval mortality critical to overall population performance, densities at a larger scale should be influenced by soil-water tables and rainfall.

Survival and Mortality factors

Mortality can be categorized into: (a) natural mortality caused by ageing and senescence, (b) mortality caused by (micro) climatic factors, (c) mortality caused by resource shortage, (d) mortality caused by cropping practices, and (e) mortality caused by biotic interactions. These categories remain arbitrary since there are all kinds of interactions. A good example is that age and physiological condition of an individual affect their potential to withstand adverse conditions. Also the developmental stages, eggs, larvae, pupae and adults have very different susceptibility for different factors. Survivorship curves derived from the field reflect the sum of all mortality factors.

Natural mortality by ageing

Under laboratory conditions mortality functions of the subadult stages often takes a simple form like a constant fraction being lost per time unit. The associated parameters are stage specific. In the life cycle juvenile stages in field often have much higher mortality rates than the adults. Maximal survival curves of adults due to senescence can also be determined in optimal laboratory conditions. Senescence of arthropods, as a physiological process, is strongly determined by temperature when there are no other constraints. Laboratory experiments can be set up to study potential effects of other factors which may or may not be important under field conditions. Such laboratory experiments can be very useful to get an idea about the ecological constraints for the different developmental stages. To a large extent, the "phenology" of the adult generation in the field is the outcome of eclosion rates from the pre-adult stage and mortality due to senescence. Moreover, such phenology curves under field conditions however, may be very erratic and variable due to catastrophic events and local conditions and hence give little information about specific factors.

Mortality caused by (micro) climate

Climate is regarded as a major factor in determining year to year fluctuations in many arthropod populations. Weather conditions may have direct effects on numbers by causing mortality under extreme conditions, or indirectly by changing food levels or food accessability.

In arable land, shelter against adverse weather conditions may be crucial for many species. In particular, in the crop-free period and during tillage, epigeic arthropods are exposed to extreme moisture and temperature conditions. Many beneficial species prefer situations where ground cover and shelter is available which creates favourable microclimatic conditions (crops, weeds, green-manure, and structure-rich soil) (Powell et al. 1985).

For the maintenance of several species, including some carabids, coccinellids, syrphids and parasitoids the presence of non-crop habitat like field margins is essential to survive during the winter (Sotherton 1984).
Farmers have no grip on weather conditions, but they can promote favourable micro climatic conditions by providing cover and shelter. Experimental studies to quantify the importance of shelter on survival of beneficial insects are lacking.

Mortality caused by limited resources

Many predatory insects like spiders and carabids can stand long periods of starvation, which certainly helps to survive in the unpredictable agricultural environment. Yet, food is generally assumed to be a key factor for population performance. For specialized beneficials like coccinellids, food levels are relatively easy to quantify. For more polyphagous species, however, the food spectrum is broad and can be extremely variable is time and space. Hence it is more difficult to define the main food sources and even more difficult to quantify food availability. Since potential food-items may also differ in nutritious quality and preferential switching occurs, this makes it very complex. The question of what predators really consume is a study in itself (Sunderland et al. 1987).

Laboratory or semi-field experiments may provide valuable information about the role of food quantity and quality on survival and reproduction (van

Dijk 1994). However, results should be interpreted with great care, in particular when it concerns polyphagous species. It seems almost impossible to provide them with a food sources which reflect field conditions.

In some cases it is possible to make indirect estimates of food conditions in the field by assessing parameters for feeding conditions of the animals like weight, respiration, stomach contents, fat contents etc. (Zhou et al. 1995). In terms of population dynamics, food is important as soon as it affects the basic parameters survival, reproduction or migration. Even though animals can stand long periods of starvation, hungry predators may be more vulnerable to weather conditions, cannibalism or to predation.

When animals are more active or disperse because of food shortage they may also face higher mortality risks. Adult stages of many species seem to have behavioural strategies to deal with food shortage, which is likely to be a common situation in agriculture.

For the larval stages of polyphagous species there is a great lack of knowledge about the role of food on growth and survival. A better insight in larval ecology is strongly needed.

Manipulation of food levels in the field is not easy, in particular because most species tend to be polyphagous, and various alternative food sources, which are often not known, vary over time. Even though higher food levels may enhance survival and hence predator numbers, predation of relevant pest organisms may be reduced due to preference for non-pest species. Several crop management practices like tillage, manuring, and weed management are likely to have an impact on food levels but quantitative studies are scarce (but see Zangger et al. 1994). Pollen feeding predators like syrphids, parasitoids and some predatory bugs may be enhanced by providing strips with pollen rich flowers. The degree to which different non-crop related food sources contribute to survival is poorly understood but some studies are available (e.g. Pemberton and Vandenberg 1993).

Mortality caused by cropping practices and habitat management

The most important cropping practices which vary between different farming systems and influence survival are pesticide-applications, soil tillage, weed management, (green) manure and cover crops, the management of field margins and presence of set aside fields.

Until recently intensive use of broad-spectrum pesticides was common in conventional agriculture. Several of these chemicals are highly toxic to many predator and parasitoid species. A vast literature exist on detrimental side effects of insecticides and other chemical treatments (Jepson 1990a).

Though many of the broad spectrum insecticides have been replaced by more specific and less toxic chemicals, pesticides should be regarded as a potential threat to predator population performance, either by reducing numbers directly or indirectly by eliminating major food resources. Toxicity tests performed in the laboratory cannot be directly extrapolated to field conditions and final effects under field conditions strongly depend on the timing and spatial scale of application. A quick recovery of local populations after application can only occur when unsprayed favourable habitats are nearby (Jepson 1990b).

Intensity of soil tillage, soil cultivation methods and timing are likely to affect species which are active on the soil surface and develop or hide in the soil during part of the year. Though many field inhabiting species will be able to withstand some degree of soil disturbance, some life stages are likely to be more vulnerable (pupae) and mortality may occur due to a sudden change in abiotic conditions, and increased predation by for example birds. The resulting structure of the soil top layer and soil surface may have profound effect on the number of shelter places or sites to settle. The number of web sites for example seem to be a limiting factor for several linyphiid species.

The presence of weeds in a crop have shown to be an important factor which offers shelter and food to a lot of beneficial arthropods. In particular parasitoids and partly phytophagous species like carabids of the genera *Harpalus* and *Amara* strongly benefit from a certain amount of weed cover (Bosch 1987). The total removal of weed either chemically or mechanically will affect conditions for survival. The beneficial effects of presence of weeds is likely to be most obvious in crops with a low soil surface cover like onions, carrots.

Likewise, green manure, cover crops and crops with a sufficient soil cover during most of the breeding season offer favourable conditions for survival in species which need some shelter and a moderate microclimate.

For part of the species inhabiting arable fields the presence of less disturbed sites adjacent to or near the cropped fields may be crucial to build up high population levels. Non-crop vegetation like field margins, uncropped strips, fallow and set-aside fields may serve as an overwintering (survival) site from which fields are recolonized in spring (e.g. Thomas 1990). In population dynamical sense these spatially separated resources are essential for the performance of the overall population. However, for many species which move from cropped into non-cropped areas and vice versa, this function of the non-cropped areas has only been proven in a qualitative sense.

Mortality due to biotic interactions
Organisms which are not used as food may affect survival of a individual either by competing for a shared resource (e.g. food, web sites) or by acting as a malentity (diseases and natural enemies). When resources are scarce – and the feeding condition of many beneficial insects in the field indicate that this is a common phenomenon – the presence of high densities of beneficials sharing the same resource leads to a decrease of survival chances of each individual (intra- and interspecific competition).

It has been suggested that cannibalism within species of carabids and staphylinids may be common and it has been shown that many arthropod predators attack other predator species even when phytophagous prey is available. Quantitative information, however, about these processes under field conditions is lacking as well as its relation with population density. Although larger predators like birds and mammals are known to feed on all kind of beneficial arthropods, also there the quantitative impact on populations is unknown.

Fragmentary and anecdotal information has been published about the occurrence of parasitic and fungal diseases in beneficials, but data so far suggest that this is not a key factor in the dynamics of most arthropod predator species in agriculture. Occasionally severe mortality due to diseases may occur.

Concluding Remarks

To be able to manipulate densities of beneficials by reducing mortality factor and hence increasing their survival, key factors should be identified which are manageable. Since many natural enemies may affect pest densities, research should be focused on a few species which are assumed to be crucial in controlling the target pest species. More detailed life history studies which identify major mortality factors are needed to obtain appropriate and quantitative information for even the most well-studied beneficial animals.

References

Boer, P.J. den 1968. Spreading of risk and the stabilization of animal numbers. *Acta Biotheoretica* 18: 165-94.

Bosch, J. 1987. Der Einfluss einiger dominanter ackerunkräuter auf Nutz- und Schadarthropoden in einem Zuckerrübenfeld. *J. Plant. Dis. Prot.* 94: 398-408

Dijk, Th.S. van 1994. On the relationship between food, reproduction and survival of 2 carabid beetles, *Calathus melanocephalus* and *Pterostichus versicolor. Ecol. Entomol.* 19: 263-70.

Jepson, P.C. & Thacker, J.R.M. 1990. Analysis of the spatial component of pesticide side-effects on non-target invertebrate populations and its relevance to hazard analysis. *Functional Ecology* 4: 349-56.

Jepson, P.C. (ed.) 1989a. *Pesticides and Non-target Invertebrates* Intercept, Andover

Jepson, P.C. 1989b. The temporal and spatial dynamics of pesticide side effects on non-target arthropods. In: Jepson P.C. (ed.) *Pesticides and Non-target Invertebrates*, Intercept, Andover, 95-127.

Jones, M.G. 1979. The abundance and reproductive activity of common Carabidae in a winter wheat crop. *Ecol.Entomol.* 4: 31-43

Luff, M.L. 1982. Population dynamics of Carabidae. *Ann. appl. Biol.* 101: 164-70

Nijs, den L.J.M.F., Booij, C.J.H., Daamen, R., Lock, C.A.M. & Noorlander, J. 1996. Can pitfall trap catches inform us about survival and mortality factors? An analysis of pitfall trap data for *Pterostichus cupreus* (Coleoptera: Carabidae) in relation to crop rotation and crop specific husbandry practices. *Acta Jutl.* this issue.

Pemberton, R.W. & Vandenberg, N.J. 1993. Extrafloral nectar feeding by ladybird beetles (Coleoptera, Coccinellidae). *Proc. Entomol. Soc. Washington* 95: 139-51.

Powell, W., Dean, G.J. & Dewar, A. 1985. The influence of weeds on polyphagous arthropod predators in winter wheat. *Crop Protection* 4: 298-312.

Sotherton, N.W. 1984. The distribution and abundance of predatory arthropods overwintering on farmland. *Ann. Appl. Biol.* 105: 423-29.

Spence, J.R. & Niemelä, J.K. 1994. Sampling carabid assemblages with pitfall traps: The madness and the method. *Can. Entomol.* 126: 881-94.

Sunderland, K.D., Crook, N.E., Stacey, D.L. & Fuller, B.I. 1987. A study of feeding by polyphagous predators on cereal aphids using Elisa and gut dissection. *J. Appl. Ecol.* 24: 907-33.

Thomas, M.B. 1990. The role of man-made grassy habitats in enhancing carabid populations in arable land. In: Stork, N.E. (ed.) *The Role of Ground Beetles in Ecological and Environmental Studies.* Intercept, Andover, 77-85.

Topping, C.J. & Sunderland. K.D. 1994. A spatial dynamics model for *Leptyphantes tenuis* (Araneae: Linyphiidae) with some simulations of the spatial and temporal effects of farming operations and land-use. *Agriculture, Ecosystems and Environment* 48: 203-17.

Wallin, H. 1987. Life cycles, reproduction and longevity of the medium sized carabid species *Pterostichus cupreus* L., *P. melanarius* Ill., *P. niger* Schaller and *Harpalus rufipes* De Geer (Coleoptera, Carabidae) inhabiting cereal fields. In: Wallin, H. 1987. Distribution, movement and reproduction of carabid beetles (Coleoptera: Carabidae) inhabiting cereal fields. Thesis, Uppsala, Sweden.

Zangger, A., Lys, J.A. & Nentwig, W. 1994. Increasing the availability of food and the reproduction of *Poecilus cupreus* in a cereal field by strip management. *Entomol. Exp. Appl.* 71: 111-20.

Zhou, X., Honek, A., Powell, W. & Carter, N. 1995. Variations in body-length, weight, fat content and survival in *Coccinella septempunctata* at different hibernation sites. *Entomol. Exp. et Appl.* 75: 99-107.

REPRODUCTION

Reproduction and survival of linyphiid spiders, with special reference to *Lepthyphantes tenuis* (Blackwall)

Sunderland, K.D.[1], Topping, C.J.[2], Ellis, S.[1], Long, S.[3], Van de Laak, S.[4], & Else, M.[5]

[1]Entomology Department, Horticulture Research International, Littlehampton, West Sussex, BN17 6LP, UK
[2]Present address: Danish Environmental Institute, Kalø Grenaavej 12, DK-8410 Rønde, Denmark
Aberdeen AB9 1UD, Scotland, UK
[3]Department of Pure & Applied Biology, University of Leeds, LS2 9JT, UK
[4]Entomology Department, Wageningen Agricultural University, Postbus 8031 6700 EH Wageningen, The Netherlands
[5]Department of Conservation Sciences, Bournemouth University, Fern Barrow, Poole, Dorset BH12 5BB, UK

Abstract

Linyphiid spiders are potentially useful for biological control, but to be able to exploit this potential (for example by manipulating their numbers in the field) it is first necessary to understand more about the factors which affect their survival and reproduction. Results of field studies on reproduction of a range of species, and of laboratory studies on reproduction and survival (mainly of *Lepthyphantes tenuis*) are summarized in this paper.

Patterns of eggsac production by *Lepthyphantes tenuis* in winter wheat indicated that breeding continues until at least November, whereas most other species terminated eggsac production a few months earlier. Mean numbers of spiderlings per eggsac in the field varied from 9 to 27, depending on species, and egg infertility was less than 10%. *L. tenuis* (n = 8045) were used in laboratory studies to determine the effect of food and temperature on development, survival and reproduction. Reproduction was compared at 15°c and 21°c. The maximum number of eggsacs per female was ten and the maximum fecundity was 191 eggs at 15°c. The reproductive performance (ie fecundity and duration of reproductive life) of wild females was greater at 15°c than 21°c. The reproductive formance of wild females was greater, at 15°c, than that of females reared in the laboratory. Reproductive performance of *L. tenuis* was greater when reared on a diet of mixed Collembola (mainly *Lepidocyrtus* spp. and *Isotoma* spp.) than on the collembolan *Folsomia candida* alone. Development and

Arthropod natural enemies in arable land · II *Survival, reproduction and enhancement*
C.J.H. Booij & L.J.M.F. den Nijs (eds.). *Acta Jutlandica* vol. 71:2 1996, pp. 81-95
 ISBN 87 7288 672 2

survival of hatchlings were compared on diets of pollen and yeast, flour mite (*Acarus siro)*, cereal aphids (*Sitobion avenae*) and various Collembola at 9°c, 12°c, 15°c and 21°c. No hatchlings survived to become adult on diets of pollen and yeast, aphids or flour mites. About half of the hatchlings survived to adult on *F. candida* at 12°c compared with only 1% at 21°c. However, 60% survived to adult on mixed Collembola at 21°c (a similar result was obtained for *Erigone promiscua* fed on *F. candida* at 21°c). Hatchlings required approximately one month to become adult at 21°c. The survival rate differed between broods within treatments and there was also much variation between individuals. For adult *L. tenuis*, females were larger than males and adults taken from the wild were larger than those reared in the laboratory. It is concluded that food quality can have a powerful influence on survival and reproduction and that there are subtle interactions between food quality and temperature. Areas requiring further work are identified and the results are discussed in relation to the ecological plasticity of *L. tenuis*.

Key words: reproductive performance, maximum fecundity, mixed diet, Collembola, food quality

Introduction

Spiders have potential as biological control agents in agriculture (Nyffeler & Benz 1987) and the Linyphiidae is the dominant family in crops in northwestern Europe (Sunderland 1987). To enable a more efficient utilization of these spiders in biological control, it will be necessary to gain an understanding of the principal factors affecting their distribution and abundance. This paper summarises results of laboratory and field studies on the reproduction and survival of linyphiid spiders found in UK cereals. Laboratory studies of the effect of food and temperature on development, survival and reproduction were done mainly with *Lepthyphantes tenuis* (Blackwall) since this species is distributed very widely, is one of the most abundant species in grass and cereals, and is likely to be especially valuable in pest control because of its production of large webs (Fraser 1982, Sunderland et al. 1986b, Topping & Sunderland 1994, Sunderland, in press). The principal foods of linyphiid spiders in arable fields appear to be aphids and Isotomidae (Collembola) (Sunderland et al. 1986a, Alderweireldt 1994). The food types used in the laboratory were constrained by the need to have a secure supply for long-term (up to one year) replicated experiments and, although some are likely to be unnatural foods for *L. tenuis*, the results illuminate principles (such as variability of performance under food stress) that may operate in the field. The investigations also provide data on the effect of food and temperature on survival and reproduction which can be used to upgrade metapopulation dynamics models of *L. tenuis* (Topping & Sunderland 1994ab).

Materials & Methods

Reproduction in the field

Linyphiid reproduction was studied in southeast England in a 4.2 ha field of winter wheat (c.v. Hobbit) in 1978, in a 6.5 ha permanent pasture in the autumn of 1978, in a 17 ha field of winter wheat (c.v. Pastiche) in 1990 and a 3 ha field of winter wheat (c.v. Riband) in 1991. Adult female spiders were collected from the field with a pooter at weekly intervals and kept individually, without food, in 9 cm diameter Petri dishes lined with moist filter paper. The dishes were returned to a ventilated box in the field and examined weekly. Females readily produced eggsacs in the dishes and the proportion doing so in the first week of incarceration was used as an index of the pattern of breeding in the field.

Although rate of eggsac production is likely to be affected by food supply, it was assumed that the effect of having no food in the dishes would not be great over a period as short as one week. This assumption was supported by results of an experiment with *Erigone* spp. (n = 72) where half the dishes in the ventilated box were supplied with five vestigial-winged fruit flies (*Drosophila melanogaster* Meigen) per dish and the rest contained no food. The percentage producing eggsacs was 14% higher where food was supplied.

After the first week, females were removed from the dishes and identified and those dishes containing eggsacs were returned to the ventilated box to be examined weekly. The number of spiderlings emerging from each eggsac was recorded and eggsacs were dissected after the spiderlings had emerged so that undeveloped eggs could be counted.

Laboratory experiments with L. tenuis

Every individual *L.tenuis* used in laboratory experiments was given a unique code which could be used to access records on that individual (e.g. date and location of capture in the field, date emerged from eggsac in the laboratory and code of mother). This system permitted an audit trail so that, for example, data on the performance of hatchlings in experiments could be grouped to enable the detection of brood effects. This was necessary because each experiment was conducted on spatially randomized groups of hatchlings from several broods.

Development & survival under laboratory conditions

Adult female *L. tenuis* were collected from the field and kept individually at room temperature in Petri dishes (as above). They were provided with an excess of vestigial-winged *Drosophila* (reared on a cornmeal, yeast and sucrose diet) three times per week. Any eggsacs produced were kept in separate Petri dishes in the absence of the mother; the filter paper was kept moist and the dishes examined three times per week to record the emergence date of hatchlings. Each development and survival experiment was initiated, in most cases, with a hundred or more hatchlings, taken from more than one *L. tenuis* eggsac (usually at least 3-4). Each hatchling was kept individually in a Petri dish (as above). The dishes were randomised and placed, in the dark, inside cardboard boxes. Most experiments were done in an insectary at 21°c (mean 21.4, SE 0.05), in incubators at 15°c (mean 15.0, SE 0.02) and 12°c (mean 11.8, SE 0.05) (plus one experiment at 27°c to 29°c), and in a cold room at 9°c (mean 9.0, SE 0.03). The humidity inside a Petri dish at 21°c, subject to the normal weekly watering regime, was 100% (mean 99.6, SE 0.1). Two experiments were done at reduced humidity (mean RH 42.8%, SE 0.23, and mean RH 35.7%, SE 0.12) in a Fisons Fi-totron 600H Growth Cabinet at 21°c (mean 21.0, SE 0.02), and in these cases Petri dishes had lids replaced with fine gauze netting and no moist filter paper was used. Foods used in the experiments were: (i) pollen (P); macerated tablets of commercial bee-collected pollen, (ii) yeast (Y); ground granules of commercial dried bread-making yeast, (iii) the springtail *Folsomia candida* (Willem) (Collembola: Isotomidae) (F) from a laboratory culture established in 1979 and reared on yeast (above), (iv) mixed Collembola (Mix Coll.); various species, mainly in the genera *Isotoma* (Isotomidae) and *Lepidocyrtus* (Entomobryidae), not reared but collected regularly from a clover and ryegrass field, (v) flour mite *Acarus siro* L. (Acari) (FM) reared on brewer's yeast and wheatgerm, (vi) the cereal aphid *Sitobion avenae* (F.) (Homoptera: Aphididae) (SA) reared on winter wheat seedlings, (vii) mixed diet (Mix diet); a combination of (i) to (vi). All dishes were examined three times per week, at which time the filter paper was re-moistened, excess food was added and deaths and moults (detected by the presence of exuviae, which were removed from the dishes) recorded. *L. tenuis* was found to be a species that is extremely reliant on a web for food capture; individuals were never observed eating food on the filter paper at the base of the Petri dish, but always in their webs which were constructed above the base of the dish. A similar conclusion was reached by Alderweireldt (1994). For this reason care was always taken to place food directly into the web.

The development and survival of *Erigone promiscua* (O.P.-Cambridge) and *Oedothorax retusus* (Westring) on *Folsomia candida* at 21°c was also measured; these species provided an interesting comparison with *L. tenuis* since they are not so web-dependent and can hunt for their prey on the surface of the filter paper in the Petri dish.

Reproduction of L. tenuis in the laboratory
Subadults from the field or from laboratory-rearing on flour mite and *F. candida* were kept individually at room temperature in Petri dishes (as above) and were provided with an excess of vestigial-winged *Drosophila* and *F. candida* three times per week. A few days after moulting to adult, the spiders were allowed to mate overnight and then females were kept individually at 15°c or 21°c in Petri dishes (as above) and were provided with an excess of *Drosophila* three times per week. Deaths and eggsac production were recorded.

Size of adult L. tenuis and of their eggs
Measurements of adult *L. tenuis* were taken under a binocular microscope at 100 times magnification using a micrometer eyepiece. Cephalothorax length (from between the posterior median eyes to the base of the cephalothorax) and width (taken at the widest point) were measured dorsally with the cephalothorax held completely flat under 70% alcohol in a glass dish. Length multiplied by width was used as an index of size.

Eggs were measured similarly in 70% alcohol at 100 times magnification. For each egg, two measurements were taken perpendicular to each other and the mean used as an index of egg size.

Results

Reproduction in the field
The pattern of change in the percentage of spiders breeding for *Erigone* spp., *Oedothorax* spp. and *Milleriana inerrans* (O.P.-Cambridge) tended to be bimodal, with peaks in spring and summer. The pattern for *Bathyphantes gracilis* (Blackwall) and *Meioneta rurestris* (C.L. Koch) was unimodal with peaks in August and October respectively. The pattern for *L. tenuis* was more complex and varied between years. *Erigone, Oedothorax* and *B.gracilis* did not breed after September, but *L.tenuis* was still producing eggsacs when sampling ended in November.

The mean number of spiderlings per eggsac ranged from 8.9 for *Erigone dentipalpis* (Wider) to 27.3 for *Oedothorax retusus* (Westring) (Table 1). There was considerable variation within the genus *Oedothorax*, but less within the genus *Erigone*. Spiderlings occasionally failed to emerge from eggsacs (due either to infertility or because dishes accidentally became too wet or too dry) but the incidence was only 6.9% of 1347 eggsacs. Undeveloped eggs were rare in eggsacs from which some spiderlings emerged; overall, undeveloped eggs constituted only 3% of potential production (spiderlings plus undeveloped eggs, n = 20,915).

Reproduction of L. tenuis in the laboratory

A comparison was made between wild and reared *L.tenuis* at 15°c (Table 2). The mean duration of reproductive life was similar, but more reared females died without producing an eggsac. Wild females tended to produce more eggsacs per female and to have a higher fecundity than reared females. The mean number of eggs per eggsac did not vary much in relation to position in the production sequence; for wild females the first eggsac contained a mean of 21.9 eggs (SE = 3.1, n = 14), the second 26.2 (SE = 2.6, n = 9), the third 21.3 (SE = 2.8, n = 7) and the fourth 20.0 (SE = 3.0, n = 7). Reproduction of wild *L. tenuis* was also measured at 21°c (Table 2). The mean duration of reproductive life was less than at 15°c and other aspects of reproductive performance were similar to reared females at 15°c. Mean number of eggs per eggsac was not affected greatly by oviposition sequence being 22.4 (SE = 1.7, n = 19) for the first eggsac, 24.7 (SE = 3.1, n = 10) for the second and 26.0 (SE = 5.3, n = 5) for the third.

For wild females at 15°c and 21°c there was no indication of change in egg size in relation to position of the eggsac in the production sequence, but there was a tendency for eggs to be smaller at 15°c (mean of mean egg size per eggsac = 0.438 mm, SE = 0.002, n = 44 eggsacs) than at 21°c (mean = 0.460 mm, SE = 0.002, n = 31).

In two of the *L. tenuis* development and survival experiments described below (12°c *F. candida*, and 21°c mixed Collembola – see Table 4) sufficient

Table 1. Mean number of spiderlings per eggsac in the field.

Species		n	Mean	SE
Erigone dentipalpis	(Wider)	45	8.9	0.5
Erigone promiscua	(O.P.-Cambridge)	74	9.5	0.5
Erigone atra	(Blackwall)	79	10.7	0.4
Oedothorax fuscus	(Blackwall)	48	13.3	0.8
Oedothorax apicatus	(Blackwall)	47	17.8	0.9
Oedothorax retusus	(Westring)	50	27.3	1.3
Lepthyphantes tenuis	(Blackwall)	324	22.2	0.6

immatures developed into adult females to allow investigation of their reproduction. On *F. candida* at 12°c, 39% of 23 females produced one eggsac and a further 17% produced two eggsacs. Only 29% of the 17 eggsacs hatched and the mean number of spiderlings per eggsac was only 5.8 (SE = 2.5, range 1-14). 31% of the 29 second generation hatchlings survived to become adult, but no viable eggsacs were produced by them. The mixed Collembola experiment at 21°c was terminated prematurely but, by 57 days from the start of the experiment, 34 immatures had become adult females, 50% of these had produced one eggsac, a further 24% two eggsacs and a further 3% three eggsacs. Twelve eggsacs had hatched yielding a mean of 20.8 spiderlings per eggsac (SE = 2.3, range = 12-34).

Development & survival under laboratory conditions

Table 3 summarises the results of experiments on the development and survival

Table 2. Reproduction of wild and reared *Lepthyphantes tenuis* in the laboratory.

	Mean	s.e.	n	range
		reared, 15 °C		
% not reproducing	36			
Duration of reproductive life (days)	58	5.2	53	
No. eggsacs per female [1]	2.4	0.6	18	1-10
Fecundity [1]	42.5	23	8	11-191
		wild, 15°C		
% not reproducing	26			
Duration of reproductive life (days)	49.2	7.8	19	
No. eggsacs per female [1]	3.3	0.7	15	1-9
Fecundity [1]	77.2	18.3	13	5-189
		wild, 21 °C		
% not reproducing	39			
Duration of reproductive life (days)	13.3	1.1	31	
No. eggsacs per female [1]	1.9	0.2	19	1-3
Fecundity [1]	42.3	5.5	19	16-101

[1] Excluding individuals that produced no eggsacs

of *L. tenuis* immatures in relation to food and temperature. At *c.* 100% RH immatures survived for a month without food at 9°c, two weeks at 21°c and one week with food at 29°c. When the humidity was reduced below 50%, immatures survived for less than one week without food at 21°c. Pollen and yeast may have been toxic to the spiders since these foods reduced mean longevity at 21°c to half that recorded in the absence of food. On the other hand a few spiders managed to moult once (moulting here, and below, refers to moulting after emergence from the eggsac) when given pollen and yeast compared with no moulting for starved spiders at 21°c. *Sitobion avenae* was a poor food and only 2% of spiders developed as far as the second moult on it. Survival was slightly better on flour mite and a few individuals survived to the third moult. No individuals fed on pollen and yeast or *S. avenae* or flour mite succeeded in developing to the adult stage (n = 802).

Survival at all temperatures was consistently better on *F. candida* alone than on *F. candida* plus pollen and yeast (Table 4). Although, in general, few individuals survived to adult on these foods, *F. candida* at 12°c proved to be an exception, with 50% survival to adult (sex ratio 1.3 male: 1 female). 60% of *L. tenuis* survived to adult (0.8 male: 1 female) on a diet of mixed Collembola at 21°c, compared with only 3% when fed a mixed diet, 1% when fed *F. candida* and 0.6% when fed *F. candida* plus pollen and yeast at the same temperature.

Table 3. Development and survival of immature *Lepthyphantes tenuis* in the laboratory.

Temperature (C)	Food [1]	Longevity n	(days) mean	SE	percentage 1	surviving 2	moults 3
21,RH<50	none	192	4.6	0.2			
21	none	60	14	0.6			
15	none	105	24	0.9			
12	none	100	21.2	1.0	1.0		
9	none	62	33.6	2.0			
21	PY	100	7.5	0.4	5.0		
27-29	FPY	206	6.8	0.4			
21	SA	288	14.1	0.7	20.8	2.1	
21	FM	104	21.2	1.3	26.9	9.6	1.0
15	FM	105	57.3	3.0	77.1	31.4	3.8
12	FM	105	82.8	6.2	15.2	1.0	
9	FM	100	24.5	2.6	1.0		

[1] Food: FPY= Folsomia candida, pollen & yeast, SA= Sitobion avenae, FM= flour mite

It took nearly two months for individuals to reach adult on a diet of *F. candida* plus pollen and yeast at 21°C, compared with about one month on the other diets. (*F. candida*, mixed diet, mixed Collembola) at the same temperature. However, with the exception of mixed Collembola, sample sizes were very small and more data are needed. Development to adult took about three months at 15°C and over four months at 9°C. The performance of *Erigone promiscua* on *F. candida* at 21°C contrasted strongly with that of *L. tenuis* under the same conditions (Table 4). Only 1% of *L. tenuis* survived to adult (taking 33 days) compared with 52% of *E. promiscua* (taking 23 days) (0.6 male: 1 female), and mean intermoult duration was twice as long for *L.tenuis*. *Oedothorax retusus*, under the same conditions, had variable intermoult periods and only 2% survived to adult, taking nearly a month to do so.

For nearly all combinations of temperature and food, mortality was greatest during the first intermoult period and considerably less during the

Table 4. Development and survival to adult, in the laboratory, of the immature *Lepthyphantes tenuis*, *Erigone promiscua* and *Oedothorax retusus*.

			Development period (days) of survivors to adult		
Temperature (C)	Food [1]	Initial n	mean	SE	% to adult
Lepthyphantes tenuis					
21	FPY	328	55.5	2.1	0.6
15	FPY	298	96.0	69.3	0.7
12	FPY	268	100.7	10.5	2.6
9	FPY	290			0.0
21	F	292	33.0	2.1	1.0
15	F	103	72.0	7.7	6.8
12	F	105	93.6	3.1	49.5
9	F	100	140.8	5.2	4.0
21	Mix diet	179	28.8	1.8	3.4
21	Mix Coll.	100	31.4	0.6	60.0
Erigone	*promiscua*				
21	F	100	22.9	0.6	52.0
Oedothorax	*retusus*				
21	F	100	26.0	1.4	2.0

[1] Foods: FPY= Folsomia candida, pollen & yeast, F= F candida, Mix Coll.= mixed Collembola (mainly Lepidocyrtus & Isotoma), Mix diet= mixed diet (all foods here plus those in Table 3).

Table 5. Examples of brood effects in laboratory experiments with *Lepthyphantes tenuis*.

Temperature (C)	Food [1]	BROOD "A" Longvity (days)				BROOD "B" Longvity (days)			
		n	% adult	Mean	SE	n	% adult	Mean	SE
21	Mix Coll.	5	0.0	3.4	0.5	5	100.0	22.2	0.8
room	FM	14	0.0	2.9	1.0	6	16.7	26.3	9.1
room	F	21	0.0	8.4	0.8	19	26.3	66.4	7.9

[1] Food: Mix Coll.= Mixed Collembola (mainly Lepidocyrtus & Isotoma), FM= flour mite, F= *Folsomia candida*.

second and third intermoult periods. Flour mite at 15°c was an exception, therebeing 46% mortality during the second intermoult period. *L. tenuis* given *F. candida* at 12°c and *E. promiscua* given *F. candida* at 21°c both experienced high mortality during the last intermoult period.

A brood effect was noted in some experiments, in that survival rate sometimes differed between broods. Some extreme examples are summarised in Table 5. The females that produced the eggsacs to initiate the mixed Collembola experiment were collected from the same field on the same date by the same method, but performance of their progeny was different. Extreme effects were also observed in second generation spiderlings (mothers had been reared on flour mite and *F. candida* at 21°c) given flour mite at room temperature or given *F. candida* at room temperature (Table 5). In addition to brood effects, in many experiments there were just one or two individuals that survived exceptionally longer than the average. Some examples of maximum longevity minus mean longevity for various experiments are; 113, 151, 166, 167, 173, 178, 193, 206, 339 days.

Table 6. Size of adult *Lepthyphantes tenuis*. Size index is cephalothorax length x width in mm.

	MALES size (mm)			FEMALES size (mm)		
Source	n	mean	SE	n	mean	SE
Wild	25	0.84	0.017	231	0.97	0.008
Mix Coll. at 21C	31	0.72	0.010	29	0.76	0.013
Other experiments	17	0.57	0.014	30	0.64	0.011

Size of adult L. tenuis
Adult males tended to be smaller than adult females (Table 6). Adults that developed from immatures given a diet of mixed Collembola at 21°c (Table 4) were smaller than wild adults. Adults that developed from immatures in all other experiments (Tables 3 & 4, plus those reared on flour mite and *F. candida* at 21°c) were smaller still (Table 6).

Discussion

Reproduction
The bimodal pattern of breeding recorded here in winter wheat for *Erigone* spp. and *Oedothorax* spp. probably resulted from there being two generations per year (De Keer & Maelfait 1987, 1988a). We recorded a unimodal pattern for *Bathyphantes gracilis*, which is also considered to be bivoltine in northern Europe (Schaefer 1976 in De Keer & Maelfait 1988a); the discrepancy might be due to very rapid development of the second generation of *B. gracilis* (this species has a cocoon development period shorter than that of the other common linyphiid species – Sunderland 1991). *L. tenuis* is also thought to have two generations per year (De Keer & Maelfait 1988a), although three peaks of hatchling density have been recorded in UK cereals (Topping & Sunderland, unpubl.) but it is not known if these represent three generations. Little information is available for the other species.

The mean number of eggs or spiderlings per cocoon was fairly constant for wild *L. tenuis* (22.2 at ambient temperatures, this study; 24.2 at ambient temperatures, Sunderland (1991); 22.3 at 21°c and 20.3 at 15°c in the laboratory), but was influenced by diet for laboratory-reared *L. tenuis*. Since efforts were made to supply food in excess in the experiments, these results suggest that reproduction is affected by qualitative aspects of diet (but see below), as was also found by Toft (in press) for *E. atra*.

The data on the percentage breeding in the field and the mean number of spiderlings per eggsac were used together with the data on female density, to construct a composite linyphiid natality curve for winter wheat in 1991 (Sunderland & Topping 1993). Natality peaked at 10-15 m^{-2} day^{-1} in late July.

Development & survival
In the field linyphiid survival is affected by abiotic as well as by biotic factors such as eggsac parasitism (Van Baarlen et al. 1994) and predation (Sunderland et al. 1994), but this paper concentrates on the effects of diet and constant

temperature on the development and survival mainly of *L. tenuis* in the laboratory.

Population performance on a pollen and yeast diet was investigated because hatchling linyphiid webs in the vegetation stratum of habitats are likely to receive wind-blown pollen, pollen falling from flowers above the web and yeasts abraded from leaf surfaces. Pollen is a highly nutritious food that promotes spiderling survival in Thomisidae (Vogelei & Greissl 1989) and Argiopidae (Smith & Mommsen 1984). Initial investigations showed that *L. tenuis* hatchlings would readily feed on pollen and yeast, but experiments established that mean longevity on these foods was less than that recorded in the absence of food, suggesting a toxic effect. Further investigations are needed to establish whether pollen and yeast are, in general, toxic to linyphiid spiderlings, or whether the results were due to a toxic component present only in the commercial formulations used here. Alternatively, since fungal spores have a deleterious effect on the performance of argiopid spiderlings (Smith & Mommsen 1984), yeast may have been the toxic component in our pollen and yeast diet. Webs of *L. tenuis* are known to catch the cereal aphid *S. avenae* and other aphid species in winter wheat fields (Sunderland et al. 1986ab, Alderweireldt 1994), but survival of hatchlings was poor on this food in the current study, in spite of care taken to supply them with predominantly first instar aphids. The value of *S. avenae* as a food for adult spiders was not investigated. Toft (in press) reported that the aphid *Rhopalosiphum padi* (L.) is also a low quality prey for *Erigone* spp. and *Pardosa* spp. At 21°c survival on a diet of *Folsomia candida* was very much lower than on a diet of mixed Collembola (mainly *Lepidocyrtus* spp. and *Isotoma* spp.). This could be due to toxicity effects (Toft, in press), nutritional effects (*F. candida* were reared on baker's yeast since 1979, whereas the mixed Collembola were wild-caught) or be a function of prey activity. Although both diets were introduced, in excess, directly into webs, three times per week, it is nevertheless possible that the quantity of *F. candida* was limiting if no further *F. candida* became caught in webs during the rest of the week (*L. tenuis* does not leave the web to feed — Alderweireldt 1994). *Lepidocyrtus* and *Isotoma* appeared to be more active than *Folsomia* and more may have entered webs during the week (but this was not quantified here).

Circumstantial evidence in favour of the activity hypothesis comes from the high survival of *Erigone promiscua* on *F. candida* in our experiments. *Erigone* are not so web-dependent as *L. tenuis* (Alderweireldt 1994) and can forage on the base of the Petri dish where *F. candida* are continuously abundant. De Keer & Maelfait (1988b) also obtained a very high survival, at 20°c, of *E.*

atra given excess Collembola (*Isotomurus palustris* (Müller) and *Isotoma viridis*. Bourlet; J.P. Maelfait pers. comm.), followed by excess *Drosophila*. *L. tenuis* survived well on a diet of *F. candida* at 12°c (but not at 21°c); this discrepancy could have been due to detoxification keeping pace with a reduced ingestion rate of toxins at the lower temperature, or result from a reduced food supply at the higher temperature if the escape rate of *F. candida* from webs is temperature-dependent. It is surprising that spiderlings fared less well on mixed diet than mixed Collembola. This may have been due to toxic components in the mixed diet that were absent from the mixed Collembola diet. In contrast, Toft (in press) found that although the aphid *R. padi* was toxic to spiders as a single species diet, small additions of this aphid to multi-species diets actually enhanced spider performance. More work is needed to elucidate the mechanisms underlying the large variation in spider performance on different diets; insights into these mechanisms could be crucial to understanding and manipulating spider populations in the field.

It is possible that *L. tenuis*, and other species of Linyphiidae, are obligate polyphages in that they may require a varied diet to satisfy all their energy and nutritional requirements whilst avoiding the accumulation of prey toxins.

The size of adult *L. tenuis* was related to diet during development. Wild-caught adults were larger than reared adults. However, since most wild-caught animals were taken from artificial web sites (holes 7 cm wide x 7 cm deep) in the field, and it is known that larger females win these sites in competition with smaller females (F. Samu et al., unpubl.) our sample of wild-caught adults is biased towards larger individuals. Size variation of adult *L. tenuis* in unbiased field samples is considerable (personal observation) and may reflect the variations in food supply during development. It is possible that differences in the size and quality of adult females account for the marked brood effects we observed in laboratory experiments. On the other hand, since similarly pronounced brood effects were also found in second-generation spiderlings where there had been a longer history of standardised conditions, it is possible that the brood effects have a genetic basis. We also found that a tiny proportion of individuals survived very much better than the average under stressful conditions of food and temperature. This may also have a genetic basis, but, whatever the mechanism, it emphasises the great ecological flexibility of this species, which may partly explain its wide distribution and abundance in European farmland.

Acknowledgments

We are grateful for funding from the UK Ministry of Agriculture Fisheries and Food, the UK Natural Environment Research Council and the Leverhulme Trust.

References

Alderweireldt, M. 1994. Prey selection and prey capture strategies of linyphiid spiders in high-input agricultural fields. *Bull. Br. arachnol. Soc.* 9: 300-8.

Baarlen, P. van, Sunderland, K.D. & Topping, C.J. 1994. Eggsac parasitism of money spiders (Araneae: Linyphiidae) in cereals, with a simple method for estimating percentage parasitism of *Erigone* spp. eggsacs by Hymenoptera. *J.appl.Ent.* 118: 217-23.

De Keer, R. & Maelfait, J.P. 1987. Life history of *Oedothorax fuscus* (Blackwall, 1834) (Araneae, Linyphiidae) in a heavily grazed pasture. *Rev. Écol. Biol. Sol.* 24: 171-85.

De Keer, R. & Maelfait, J.P. 1988a. Observations on the life cycle of *Erigone atra* (Araneae, Erigoninae) in a heavily grazed pasture. *Pedobiologia* 32: 201-12.

De Keer, R. & Maelfait, J.P. 1988b. Laboratory observations on the development and reproduction of *Erigone atra* Blackwell, 1833 (Araneae, Linyphiidae). *Bull. Br. arachnol. Soc.* 7: 237-42.

Fraser, A.M. 1982. The role of spiders in determining cereal aphid numbers. Ph.D. thesis, University of East Anglia.

Nyffeler, M. & Benz, G. 1987. Spiders in natural pest control: A review. *J. Appl. Ent.* 103: 321-39.

Smith, R.B. & Mommsen, T.P. 1984. Pollen feeding in an orb-weaving spider. *Science* 226: 1330-32.

Sunderland, K.D. 1987. Spiders and cereal aphids in Europe. *Bull. SROP/WPRS* 1987/X/1: 82-102.

Sunderland, K.D. 1991. The ecology of spiders in cereals. *Proc. 6th Int. Symp. Pests and Diseases of Small Grain Cereals and Maize*, Halle/Saale, Germany, 1: 269-80.

Sunderland, K.D. (in press). Studies on the population ecology of the spider *Lepthyphantes tenuis* (Araneae: Linyphiidae) in cereals. *Bull. SROP/WPRS.*

Sunderland, K.D., Fraser, A.M. & Dixon, A.F.G. 1986a. Distribution of linyphiid spiders in relation to capture of prey in cereal fields. *Pedobiologia* 29: 367-75.

Sunderland, K.D., Fraser, A.M. & Dixon, A.F.G. 1986b. Field and laboratory studies on money spiders (Linyphiidae) as predators of cereal aphids. *J.appl.Ecol.* 23: 433-47.

Sunderland, K.D., Ellis, S.J., Weiss, Topping, C.J. & Long, S.J. 1994. The effects of polyphagous predators on spiders and mites in cereal fields. *Proc. BCPC Conf. - Pests & Diseases*, 1151-56.

Sunderland, K.D. & Topping, C.J. 1993. The spatial dynamics of linyphiid spiders in winter wheat. *Mem. Queensl. Mus.* 33: 639-44.

Toft, S. (in press). The value of aphids (*Rhopalosiphum padi*) as food for cereal spiders (*Erigone atra* and *Pardosa* spp.). *J.appl.Ecol.*

Topping, C.J. & Sunderland, K.D. 1994a. A spatial population dynamics model for *Lepthyphantes tenuis* (Araneae: Linyphiidae) with some simulations of the spatial and temporal effects of farming operations and land-use. *Agr. Ecos. & Env.* 48: 203-17.

Topping, C.J. & Sunderland, K.D. 1994b. The potential influence of set-aside on populations of *Lepthyphantes tenuis* (Araneae: Linyphiidae) in the agroecosystem. *Asp. Appl. Biol.* 40: 225-28.

Vogelei, A. & Greissl, R. 1989. Survival strategies of the crab spider *Thomisius onustus* Walckenaer 1806 (Chelicerata, Arachnida, Thomisidae). *Oecologia* 80: 513-15.

The foraging efficiency of the parasitoid *Diaeretiella rapae* (Hymenoptera: Braconidae) in relation to the spatial distribution of aphids on plants

Wilf Powell & Andrea Nickless

Entomology & Nematology Department, IACR-Rothamsted, Harpenden, Herts., AL5 2JQ, UK

Abstract
In laboratory experiments the foraging success of the aphid parasitoid *Diaeretiella rapae* was greater when its aphid hosts were scattered individually on chinese cabbage plants than when they were aggregated. This result is contrary to that predicted by theoretical models and reasons for this are discussed. In particular, the spatial scale at which host aggregation is defined is very important when investigating parasitoid foraging behaviour in relation to host distribution. Our results suggest that *D. rapae* is an efficient forager at low host densities when aphids are dispersed on individual plants, but infested plants may be aggregated within crops.

Key words: aphid parasitoid, spatial distribution, foraging, *Diaeretiella rapae*, Braconidae, *Myzus persicae, Lipaphis erysimi*

Introduction

The distribution of aphids can be considered at a number of different spatial scales: (1) aphids on individual plants, (2) infested plants within food plant patches, (3) food plants within habitats, and (4) habitats within the landscape. At the individual plant level, the distribution of immigrating alates may be random during colonisation but most aphids gradually become more aggregated as colonies develop. The degree of within-plant aggregation varies amongst aphid species, depending upon the tendency for newborn nymphs to settle close to their mothers and the dispersal behaviour of reproducing adults, but many species become highly aggregated. The foraging efficiency of aphid parasitoids

Arthropod natural enemies in arable land · II *Survival, reproduction and enhancement*
C.J.H. Booij & L.J.M.F. den Nijs (eds.). *Acta Jutlandica* vol. 71:2 1996, pp. 97-106
 ISBN 87 7288 672 2

could be affected by the spatial distribution of its host at the plant level. In theory, aphid parasitoid foraging behaviour should have evolved to ensure optimal success, as measured by numbers of offspring surviving to the next generation, when hosts are aggregated. However, if parasitoids are to be successfully employed as biocontrol agents they must decrease aphid numbers or significantly reduce the growth rate of pest populations in the early stages of an infestation when the aphids are usually less aggregated.

Mathematical models based on parasitoid behaviour patterns, particularly time allocation in host patches, have predicted that parasitism should increase with increasing host aggregation and increasing mean host densities (Waage 1979, Wellings 1993). In the case of aphid parasitoids this prediction needs to be considered at the scale of the individual plant. Hafez (1961) observed foraging females of the aphid parasitoid *Diaeretiella rapae* (McIntosh) searching on individual cauliflower leaves, which were heavily infested with the cabbage aphid *Brevicoryne brassicae* (L.). He concluded that the parasitoids searched at random and only detected the aphids from a distance of a few millimetres and, therefore, were more likely to encounter a colony rather than an isolated aphid because of the greater space occupied by the former. Furthermore, after successfully ovipositing in a host, parasitoids often moved around within a short distance of the attacked aphid, thereby increasing their chances of encountering other individuals in the same colony. However, Wellings (1993) found no systematic trends in the foraging success of the aphid parasitoid *Aphidius ervi* Haliday when its host *Acyrthosiphon kondoi* Shinji was presented in various combinations of density and levels of aggregation.

We measured the foraging success of *D. rapae* when searching individual Chinese cabbage plants infested with aphids at a constant density but with contrasting spatial distributions.

Materials and Methods

D. rapae females used in the experiments were obtained from laboratory populations maintained on *Myzus persicae* (Sulz.) and *Lipaphis erysimi* (Kalt.) feeding on Chinese cabbage plants. In all experiments aphids were presented to the parasitoids on 3 to 4 week old Chinese cabbage plants which had five or six leaves and were growing in individual 130 mm plastic pots. Ten aphids were placed on each plant and left to settle for 24hrs before the start of each experiment. Aphids were placed on the underside of leaves and either randomly spaced as individuals over the whole plant or aggregated together in a single colony on one of the larger leaves. During the settling period the aggregated aphids were confined within a simple clip cage (25 mm diameter) to prevent dispersal. The aphid-infested plants were placed in cages into which female parasitoids were released at a density of two per plant. Parasitoids remaining alive after 24hrs were removed and the plants left in the cages until parasitised aphids mummified when the number of mummies per plant was recorded. All experiments were done in a laboratory at room temperature.

Experiment 1

Five plants were placed in each of two large cages (152 cm x 75 cm x 113 cm high), consisting of a wooden frame with fine mesh netting sides and back and clear perspex doors at the front. Ten adult *L. erysimi* were placed on each plant, spaced on the plants in one cage and aggregated on those in the other. Because reproduction occurred during the 24hr settling period, the total number of aphids present on each plant was counted immediately prior to parasitoid release. The experiment was performed twice.

Experiment 2

The procedure was the same as in experiment 1 except that aphid nymphs (instars 2-4) were used instead of adults. This experiment was performed three times.

Experiment 3

The procedure was the same as in Experiment 2 except that *M. persicae* nymphs were used instead of *L. erysimi.* This experiment was performed only once.

Experiment 4

In this experiment cages measuring 48 cm x 48 cm x 63 cm high were used. Each cage contained a single plant infested with ten *M. persicae* nymphs, either individually spaced or aggregated. Six cages were used and arranged in three replicate pairs, one with spaced and one with aggregated aphids in each pair. This experiment was performed four times. The design incorporated true replication of treatments, allowing the data to be analysed using an analysis of variance.

Table 1. Foraging success of *D. rapae* attacking spaced and aggregated *L. erysimi* on Chinese cabbage plants in Experiment 1 (see text for details).

	Experiment Run No.1	Aphids Spaced	Aphids Aggregated
No. Aphids available	1	98	96
	2	132	145
	total	230	241
No. Mummies Formed	1	44	31
	2	60	23
	total	104	54
% Parasitism	1	45	32
	2	46	16
	Overall	45	22

Table 2. Foraging success of *D. rapae* attacking spaced and aggregated *L. erysimi* on Chinese cabbage plants in Experiment 2 (see text for details).

	Experiment Run No.1	Aphids Spaced	Aphids Aggregated
No. Aphids available	1	60	54
	2	58	59
	3	59	60
	total	177	173
No. Mummies Formed	1	24	0
	2	27	6
	3	5	0
	total	56	6
% Parasitism	1	40	0
	2	47	10
	3	9	0
	Overall	32	3

Results

Experiment 1

Due to reproduction during the aphid settling period the number of hosts available for attack by the parasitoids was considerably greater than the initial ten per plant, but was very similar in the two cages (Table 1). In both runs of the experiment more mummies were formed in the cage with spaced aphids than in the cage with aggregated aphids, so that over the whole experiment the percentage parasitism for the former spatial distribution pattern was twice that for the latter (Table 1).

Experiment 2

During the aphid settling period a few final instar nymphs became adults and produced offspring, but the number available for parasitoid attack remained similar in both cages (Table 2). In all three runs of the experiment more mummies were formed in the cage with spaced aphids than in the cage with aggregated aphids (Table 2). In two repeats of the experiment no mummies at all were formed in the aggregated aphid cage. Overall, the percentage parasitism was much greater when aphids were spaced than when they were aggregated (Table 2).

Experiment 3

In this experiment *M. persicae* was used instead of *L. erysimi*. Although the number of aphids available was very similar in the two cages, no mummies were formed in the aggregated aphid cage compared with 18 in the spaced aphid cage (Table 3).

Table 3. Foraging success of *D. rapae* attacking spaced and aggregated *M. persicae* on Chinese cabbage plants in Experiment 3 (see text for details).

	Aphids Spaced	Aphids Aggregated
No. Aphids Available	55	57
No. Mummies Formed	18	0
% Parasitism	33	0

Table 4. Foraging success of *D. rapae* attacking spaced and aggregated *M. persicae* on Chinese cabbage plants in Experiment 4 (see text for details).

	Experiment Run No.1	Aphids Spaced	Aphids Aggregated
No. Aphids available	1	33	31
	2	35	38
	3	28	33
	4	32	33
	total	128	135
No. Mummies Formed	1	5	0
	2	15	7
	3	3	7
	4	12	2
	total	35	16
% Parasitism	1	15	0
	2	43	18
	3	11	21
	4	38	6
	Overall	27	12

Experiment 4

In three of the four runs of this experiment more mummies were formed in the spaced aphid cages than in those with aggregated aphids (Table 4). Overall, twice as many mummies were formed in the spaced aphid cages than in the aggregated ones and this difference was statistically significant at the 5% level.

Discussion

In contrast to model predictions (Wellings 1993), the foraging success of *D. rapae* was greater in all four experiments when the spatial distribution of its host consisted of scattered individuals on plants rather than aggregated colonies, even though observations of the foraging behaviour of *D. rapae* on single leaves heavily infested with hosts suggested greater efficiency with increased host aggregation (Hafez 1961). Thus, our results were contrary to those expected and, whereas Wellings (1993) found no systematic trends relating to levels of aggregation in parasitoid foraging success, we recorded a trend towards greater success at lower levels of host aggregation. However, there are a number of factors which may influence the relationship between aphid spatial distribution and parasitoid foraging success and these need to be considered in the context of our experiments. These include the spatial scale at which aggregation is defined, the behaviour of aphids during parasitoid encounters, parasitoid behaviour and the density of both parasitoids and hosts.

Our experiments only examined aphid spatial distribution at the level of the individual plant. *D. rapae* forages on plants by walking in a well-defined search pattern which is influenced by encounters with honeydew deposits and with hosts (Ayal 1987). Foraging at higher levels in the spatial scale hierarchy involves different behaviour patterns. For example, during host habitat location *D. rapae* is known to respond to plant volatiles emanating from the brassica food plants of its hosts (Read et al. 1970), and there is recent evidence that some aphid parasitoids detect changes in the volatile spectrum emitted by plants in response to aphid feeding damage (Guerrieri et al. 1993, Grasswitz & Paine 1993). Therefore, the spatial distribution of hosts within plant or habitat patches could have a very different effect on foraging success than does their distribution on individual plants. *Diaeretiella rapae* females found small and large patches of host food-plants at equal rates but spent more time foraging on larger patches and on patches with high host densities (Sheehan & Shelton 1989). Obviously, the effects of host aggregation on parasitoid foraging success needs to be investigated by means of manipulative experiments at a range of spatial scales, including manipulation of the spatial distribution of host-infested plants within plant patches. When manipulating host patchiness it is also important to consider the parasitoids perception of spatial heterogeneity which may not be based on visual criteria but on olfactory ones.

Aphid behaviour during parasitoid attack may partly explain the poorer foraging success recorded when aphids were aggregated than when they were scattered on plants in our experiments. Parasitoid attacks resulted in much aphid movement within aggregations of both *M. persicae* and *L. erysimi* and, in the case of the latter, many aphids dropped from the leaf. This defensive behaviour, probably stimulated by the release of an aphid alarm pheromone (Pickett & Griffiths 1980), greatly reduced the parasitoids chances of achieving more than one successful oviposition within each colony. However, when aphids were scattered singly throughout the plant, attacks on isolated aphids did not disturb aphids on other leaves.

Parasitoid oviposition behaviour differs between species and must be considered when evaluating the importance of host aggregation for foraging success. Some aphid parasitoids attack their hosts slowly, minimising disturbance, and work their way gradually through a dense colony until most individuals have been attacked (Stary 1970). Others oviposit rapidly, attacking only a few aphids in a colony before moving away to rest or to search for another colony. *Diaeretiella rapae* belongs to the latter category and often leaves the plant altogether to rest between bouts of foraging (Hafez 1961). Thus, in our cages the parasitoids were frequently leaving plants and then recommencing their foraging after a rest period. At the low aphid densities used in the experiments this behaviour probably resulted in a greater chance of host encounters on plants with randomly spaced aphids than where they were aggregated.

The time spent foraging on a plant before a parasitoid leaves that plant depends upon a number of factors including the presence of foraging cues, such as honeydew deposits, and the rate of successful encounters with hosts (Waage 1979, Gardner & Dixon 1985, Hågvar & Hofsvang 1989, Sheehan & Shelton 1989, Budenberg et al. 1992). In our experiments, the plants were infested with few aphids only 24hrs before the parasitoids were released. Therefore, there were relatively few honeydew deposits present on leaves and this probably influenced the within-plant foraging pattern of the parasitoids (Ayal 1987), a problem also noted by Wellings (1993). However, honeydew deposits are also likely to be sparse on crop plants during the early stages of aphid infestation, when high foraging success rates are required for efficient pest control.

Another factor which requires further investigation is the overall density of aphids on plants. When *D. rapae* attacked *L. erysimi* on individual mustard leaves, encounter rates with hosts and oviposition attacks increased with increasing aphid densities (Pandey et al. 1984). Each plant in our experiments was initially infested with only ten aphids and in the aggregated treatment these formed a single small colony on one leaf. Thus the chances of a foraging parasitoid encountering at least one aphid before leaving the plant were much higher in the spaced treatment, in which every leaf had either one or two aphids present. Differences in encounter probabilities between the two treatments could be reduced by providing higher overall aphid densities with small colonies present on several leaves, and by increasing honeydew deposits on leaves below colonies.

Finally, the overall density of foraging parasitoids may influence the foraging success of individual females. Intraspecific interference between foraging female *Nasonia vitripennis* (Walk.) on highly aggregated patches of its host, *Calliphora* fly pupae, was proposed to explain decreasing parasitoid efficiency with increasing host aggregation in laboratory arenas (Jones & Turner 1987). As only two females per plant were used, this is unlikely to have been a significant factor in our experiments.

Since effective control often depends upon high rates of parasitism at the early stages of an aphid infestation (Wratten & Powell 1991), foraging efficiency at low aphid densities is important when evaluating parasitoids as biological control agents. At this time, aphids are usually dispersed on individual plants but may be aggregated with respect to infested plants within the crop field. The results of our experiments suggest that *D. rapae* is an efficient forager at low levels of aphid aggregation when it is searching infested plants. However, its efficiency in relation to the spatial distribution of aphids at greater spatial scales needs further investigation.

Acknowledgements

The authors thank Miss Ann Wright for assistance with maintenance of the aphid and parasitoid cultures.

References

Ayal, Y. 1987. The foraging strategy of *Diaeretiella rapae* 1. The concept of the elementary unit of foraging. *J. Anim. Ecol.* 56: 1057-68.

Budenberg, W.J., Powell, W. & Clark, S.J. 1992. The influence of aphids and honeydew on the leaving rate of searching aphid parasitoids from wheat plants. *Entomol. exp. appl.* 63: 259-64.

Gardner, S.M. & Dixon, A.F.G. 1985. Plant structure and foraging success of *Aphidius rhopalosiphi* (Hymenoptera: Aphidiidae). *Ecol. Entomol.* 10: 171-79.

Grasswitz, T.R. & Paine, T.D. 1993. Effect of experience on in-flight orientation to host-associated cues in the generalist parasitoid *Lysiphlebus testaceipes*. *Entomol. exp. appl.* 68: 219-29.

Guerrieri, E., Pennacchio, F. & Tremblay, E. 1993. Flight behaviour of the aphid parasitoid *Aphidius ervi* (Hymenoptera: Braconidae) in response to plant and host volatiles. Eur. *J. Entomol.* 90: 415-22.

Hafez, M. 1961. Seasonal fluctuations of population density of the cabbage aphid, *Brevicoryne brassicae* (L.), in the Netherlands, and the role of its parasite *Aphidius (Diaeretiella) rapae* (Curtis). *Tijdschrift over plantenziekten* 67: 445-548.

Hågvar, E.B. & Hofsvang, T. 1989. Effect of honeydew and hosts on plant colonisation by the aphid parasitoid *Ephedrus cerasicola*. *Entomophaga* 34: 495-501.

Jones, T.H. & Turner, B.D. 1987. The effect of host spatial distribution on patterns of parasitism by *Nasonia vitripennis*. *Entomol. exp. appl.* 44: 169-75.

Pandey, K.P., Singh, R. & Tripathi, C.P.M. 1984. Functional response of *Diaeretiella rapae* (McIntosh) (Hym., Aphidiidae), a parasitoid of the mustard aphid *Lipaphis erysimi* Kalt. (Hom., Aphididae). *Z. ang. Ent.* 98: 321-27.

Pickett, J.A. & Griffiths, D.C. 1980. Composition of aphid alarm pheromones. *J. Chem. Ecol.* 6: 349-60.

Read, D.P., Feeny, P.P. & Root, R.B. 1970. Habitat selection by the aphid parasite *Diaeretiella rapae* (Hymenoptera: Braconidae) and hyperparasite *Charips brassicae* (Hymenoptera: Cynipidae). *Can. Entomol.* 102: 1567-78.

Sheehan, W. & Shelton, A.M. 1989. Parasitoid response to concentration of herbivore food plants: finding and leaving plants. *Ecology* 70: 993-98.

Starý, P. 1970. Biology of Aphid Parasites (Hymenoptera: Aphidiidae) with Respect to Integrated Control. Dr. W. Junk, The Hague.

Waage, J.K. 1979. Foraging for patchily distributed hosts by the parasitoid, *Nemeritis canescens*. *J. Anim. Ecol.* 48: 353-71.

Wellings, P.W. 1993. Foraging behaviour in aphid parasitoids: Spatial scale and resource assessment. *Eur. J. Entomol.* 90: 377-82.

Wratten, S.D. & Powell, W. 1991. Cereal aphids and their natural enemies. In: Firbank, L.G., Carter, N., Darbyshire, J.F. & Potts, G.R. (eds.) *The Ecology of Temperate Cereal Fields*. 32nd Symposium of the British Ecological Society, Blackwell, Oxford. 233-57.

Indicators of prey quality for arthropod predators

Søren Toft

Department of Zoology, University of Aarhus, Building 135
DK-8000 Aarhus C, Denmark

Abstract
Prey quality is defined as a prey type's potential contribution to the fitness of the predator. The optimal nutritional composition of prey must be inferred from studies of performance, with total fitness as the ultimate criterion. Examples are given in which conflicting conclusions regarding the quality of prey species are obtained when single life history parameters are used as substitutes for total fitness. A graphical model is presented relating predator fitness to the maximal consumption rate of possible prey types, distinguishing chemically defended prey (reduced consumption and fitness) and nutrient deficient prey (compensatory consumption, reduced fitness). It seems that all prey types are nutritionally incomplete and most predators can improve their fitness by choosing a mixed diet. The quality of one prey type to a predator therefore depends on its recent feeding history. Thus, prey quality is a dynamic concept.

Key words: diets, fitness, prey quality, nutrient composition, prey defence

Introduction

It is possible to distinguish two types of food quality criteria used in research on animal foraging, pertaining to 1) the composition of the food, and 2) the performance of the animals, respectively. In general specific requirements regarding the composition of the food are very little known as far as arthropod predators are concerned (Hagen 1987, Riechert & Harp 1987), and it is only through studies of the animals' performance that we can asses these requirements.

Most theory and models of animal food selection assumes that food can be evaluated, or at least ranked, according to its value to the animal (Charnov 1976). It is also implicit, at least as far as predators are concerned,

Arthropod natural enemies in arable land · II *Survival, reproduction and enhancement*
C.J.H. Booij & L.J.M.F. den Nijs (eds.). *Acta Jutlandica* vol. 71:2 1996, pp. 107-116
 ISBN 87 7288 672 2

that the animal itself is able to do the ranking. Thus, by its choosiness, the predator has influence on the fulfilment of its needs and, by consequence, on its performance. Since fitness is a measure of total individual performance (or the reward for actual performance), the quality of a prey type may tentatively be defined as its potential contribution to the fitness of the animal.

Characteristics of the prey which influence its quality as food include a) its energy content, b) nutrient constituents, and c) defensive substances. Optimal foraging theory traditionally measures prey quality as energy gained per unit handling time (Pyke 1984, Stephens & Krebs 1986). As an indicator of food quality this falls short because it assumes all prey types to be equal in nutritional composition (Pulliam 1975, Stephens & Krebs 1986). Animals, including insects and spiders, are thought to be able to select prey items in a mixed diet that optimizes total nutritional intake, a phenomenon called dietary self-selection (Waldbauer & Friedman 1991). The species' own composition of nutrients is then often assumed to be a criterion for optimal nutritional composition for the prey (cf. Greenstone 1979, regarding essential amino acids) though this assumption may never have been tested. Defensive mechanisms of the prey reduce its value as food for the predator. A hard exoskeleton may reduce the energy obtained per unit handling time, while chemical defence (deterrents and toxins) may make otherwise rich and perfect prey unsuitable. The three food constituents may vary independently of each other from prey type to prey type, making analysis of optimal prey selection in the wild a complicated undertaking.

Any aspect of the life history of animals as well as several behavioural criteria can be used to evaluate prey quality. Among the latter, simple feeding rate (one prey experiments) or preference (two- or multi-prey experiments) are the most commonly used (Krebs 1989). As a substitute for total fitness any life history parameter may serve: growth rate, survival, reproduction (fecundity, egg size) etc. None of these are perfect measures of total fitness; examples given below will show that various life history parameters may give conflicting inferences regarding the quality of a certain prey type. This creates a need for a more dynamic concept of prey quality than the one given above.

Food uptake and prey quality

Seen from the predator's point of view it is very likely that the three constituents of the food (energy, nutrients, defensive substances) also interact with each other. For example, a prey type may offer high levels of energy and nutrients but if it also contains deterrent and/or toxic chemicals, these benefits may be inaccessible to the predator. Eisner & Eisner (1991) found that a wolf spider readily consumed adults and larvae of a moth species that had been raised on artificial diets containing no alkaloids, but totally neglected the same prey if alkaloids had been added to the diet. Prey animals collected in the field had developed on alkaloid-containing host plants, and were similarly rejected.

Aphids may have similar properties when used as prey by polyphagous predators. In a simple preference experiment wolf spiders *Pardosa prativaga* were allowed to choose between aphids *Rhopalosiphum padi* taken from the primary host *Prunus padus* and from a laboratory culture on wheat seedlings, representing the secondary host. As the aphids are differently coloured on the two hosts, they are easy to distinguish during the test. Since they also reach a much larger size on the primary host, individuals of similar size were staged together. The results showed a clear preference for aphids from the primary host: in four hours 18 spiders had eaten 32 and 11 aphids, respectively (Wilcoxon matched-pairs test, $p<0.001$).

Toft (in press) measured the food uptake of the wolf spider *Pardosa prativaga* as a function of hunger state. Groups of spiders were starved for 0, 3, 7, and 14 days, respectively. For each starvation period some spiders were then fed fruit flies *Drosophila melanogaster*, whilst others were fed aphids *Rhopalosiphum padi*. As expected the fruit fly series revealed a dramatic increase of consumption as a result of starvation (from c. 1 fly/24hrs by satiated spiders to >9 flies/24hrs after 14 days starvation). However, the group fed aphids consumed very little and most importantly; consumption was independent of hunger state. In biomass terms the intake of aphids after two weeks starvation was only 1/40 that of fruit flies. One interpretation of these results is that the aphids are defended against spider predation by some (as yet unknown but probably chemical) mechanism. This defence seems to create an upper limit to the amount of aphids that a spider can consume, no matter how hungry it becomes. Further, there was no evidence that they killed other available aphids without consuming them as is often the case, when fruit flies are offered to a hungry wolf spider (wasteful killing, Riechert & Lockley 1984).

A very similar situation exists in the carabid beetle *Agonum dorsale* (Bilde & Toft 1994). Food uptake in relation to starvation (0 and 7 days only) was measured as above; however, additional groups could choose between fruit flies and aphids. As with the spiders, when given pure diets, consumption of aphids was the same, whether the beetles were starved or not;

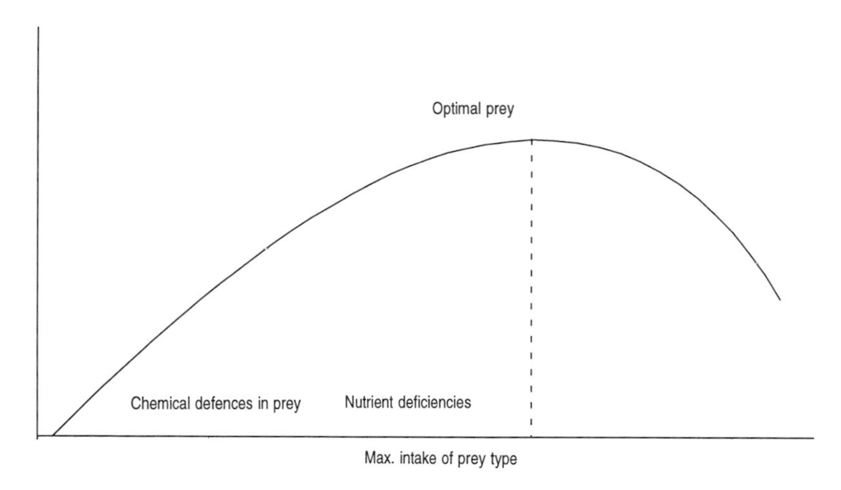

Fig. 1. Hypothetical relationship between predator potential fitness and maximal uptake of various types of prey.

however, with mixed diets (fruit flies and aphids) starved animals ate more aphids than unstarved, but still less than in the pure diet treatment. There was no indication of higher tolerance to aphids when these were given together with preferred prey; the idea of a tolerance limit indicated by the one-prey experiment was not violated in the mixed-diet situation.

If energy, nutrients and defensive substances interact with each other, how can we think of the relationship between the three in a combined way? I propose the following (admittedly hypothetical) relationship between food uptake (i.e. biomass) and fitness as a possible result of the interaction with and between the two other constituents (Fig. 1):

On the x-axis every point denotes the maximum amount the predator can eat of the various kinds of prey in a certain period of time (say 24hrs), i.e. tolerance limits to all possible prey types, with every point reflecting a different prey species. A truly perfect prey, i.e. one that contains all the energy and nutrients the predator needs in optimal composition and devoid of unpleasant chemicals (maybe non-existent), should bring the predator the highest possible fitness. Prey with defensive chemicals, which are supposedly difficult for the predator to handle physiologically, are eaten in smaller amounts, because there is either a physiological limitation to the handling of the defensive chemicals, or a point where the metabolic costs of handling the food would surpass the energy obtained; whatever the reason this is denoted the "tolerance limit". The aphids of the above examples should all be placed somewhere on the left side of the optimum. Other aphids, some collembola, sciarid flies, thrips should be placed likewise, as far as spiders are concerned (Toft, in press, A. Eberhard, unpubl.). More interestingly, the model assumes that prey types which are deficient in some nutrients that are essential to the predator, but which do not contain defensive chemicals, may be eaten in even larger amounts than the optimal prey. This is the effect we expect if the predator compensates for low nutritional quality by eating more; if combined with lowered assimilation rates, while retaining selectively the nutrients that are present in limited amounts, a more optimal composition of assimilated food is obtained. Compensatory feeding on low quality food is well known in herbivores (Haukioja et al. 1991). *Drosophila*, which we are using extensively in our experiments as a "control prey" because most of the polyphagous predators seem to have no problems with handling them and can eat large amounts, is probably positioned somewhere to the right of the optimum, as will be explained below. Van Dijk (1995) presents very convincing evidence that blowfly larvae are eaten in excessive amounts by the carabid beetle *Bembidion tetracolum* though it performs badly on this food, while the eggs of another carabid (*Pterosticus cupreus*) seem to be close to the optimum for *B. tetracolum*. Experiments with juveniles of the sheet-web spider *Oedothorax apicatus* (S. Toft, unpubl.) showed that soft-bodied soil mites were highly palatable, but a diet of these mites gave a low success for completion of development.

The relationship shown in Fig. 1 illustrates the difficulties of interpreting quantitative food uptake as an indicator of food quality, although it is a basic assumption of many types of preference tests. This may explain why several experiments have found no simple relationship between preference and performance (e.g. Chapman & Sword 1994).

Predator performance in prey quality tests

This section briefly describes an experiment which illustrates some of the problems in this approach. The results have been published in detail elsewhere (Toft, in press). Several parameters, including egg production and hatching success, were measured for groups of the spider *Erigone atra* on four different diets of fruit flies and aphids *Rhopalosiphum padi*. The four diets compared were: (1) shortage diet of *Drosophila* alone (three flies per week), (2) shortage diet of *Drosophila* supplemented with aphids *ad lib*, (3) rich *Drosophila* diet (three flies per day), and (4) aphids *ad lib*. Each group had 17-19 pairs of spiders.

The experiment was started with adult females and males collected at emergence from hibernation, before reproduction had started in the field; these animals probably had received a naturally varied diet before the experiment. It is important to realize that with an experimental design such as this, starting with animals from the wild and keeping them on diets which are all insufficient in one respect or another, it is inevitable that dietary deficiencies will tend to build up body deficiencies in the experimental animals. Thus, treatment effects will be small at the start of the experiments, but will tend to increase with time as the experiment progresses. In statistical terms this means that treatment effects should actually be looked for in the treatment*time (here sac number) interaction term.

Table 1 summarizes the results. As expected, the rich *Drosophila* diet boosted egg production, compared to the two shortage diets. On the three diets containing *Drosophila* the females produced roughly the same number of eggsacs, but the number of eggs per sac, and consequently total egg production, was far higher on the rich *Drosophila* diet. Thus by a fecundity criterion *Drosophila* would be evaluated highly. On the pure aphid diet reproduction was heavily impaired: already after the first eggsac the number of eggs per sac decreased drastically, and few females succeeded in producing a third and a fourth sac. Thus, aphids should be valued very low, as already indicated by the feeding experiments above.

Table 1. Egg production and reproductive success (avg. ± SD) of *Erigone atra* females on four diets of *Drosophila melanogaster* (=Dros.) and aphids *Rhopalosiphum padi*. Sample size varies between 17 and 19.

	3 Dros./week	3 Dros./week + Aphids	3 Dros./day	Aphids
No. eggsacs/♀	7.45 ± 2.42	8.69 ± 2.89	8.32 ± 3.57	2.83 ± 0.79
range	4-13	3-13	2-14	2-4
No. eggs/♀	62.8 ± 19.6	79.1 ± 31.9	120.3 ± 57.9	22.1 ± 8.2
range	29-101	30-144	26-227	11-39
No. young/♀	37.5 ± 12.1	53.6 ± 20.5	40.8 ± 16.0	20.1 ± 8.2
range	18-67	24-96	0-75	3-33

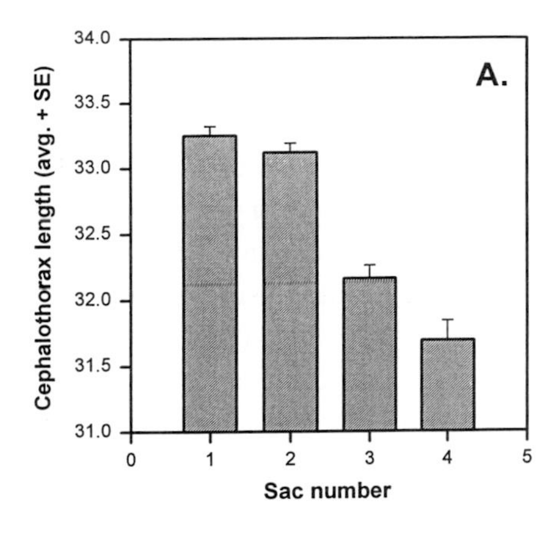

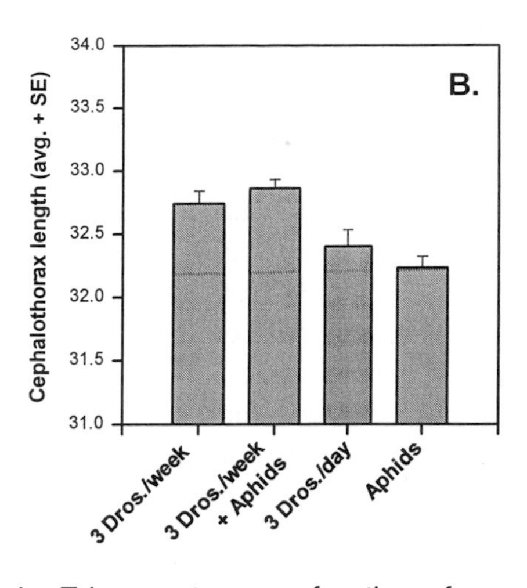

Fig. 2. Hatchling size (micrometer units) in *Erigone atra* as a function of eggsac number (A) and diet of the parents (B) in a laboratory experiment with four diets of *Drosophila melanogaster* and the aphid *Rhopalosiphum padi.*

Hatching success, however, gave a very different picture. Already after the first eggsac produced by the females, a clear picture emerged: the rich *Drosophila* diet gave a very low success, significantly lower than the shortage *Drosophila* diet (Table 1); by far the best hatching success was obtained with the shortage of *Drosophila* supplemented with aphids. As a result of this, looking at total breeding success, i.e. the number of living young produced per female on the different diets, the aphid supplemented *Drosophila* diet turned out to be the superior, the rich and the shortage *Drosophila* diets were about equal, and the pure aphid diet bad. In this situation any statement about the food quality of the two prey types will be contradictive. Aphids are low quality alone, but gave a very significant improvement to the *Drosophila* diet. *Drosophila* alone is also inferior, since they produce eggs of very low quality. With respect to hatching success alone, the shortage diet of *Drosophila* was actually much better than the rich, though evaluated by reproductive success they were about equal.

It is perhaps surprising that hatching success is so much better on a shortage diet of fruit flies than on an abundance diet. Two mechanisms may be suggested as explanations: One is that the spiders actually obtain a more varied nutrition from the food, because a shortage diet forces them to exploit the prey completely, as suggested by Toft (in press). Support for this idea is presented by Cohen (1995) who found that the proportion of the different nutrients (proteins, glycogen and lipids) extracted by a heteropteran from a prey changed during the handling period. Another may be that food limited animals put more resources into each single offspring than overfed ones. I will comment further on the latter hypothesis below.

Assuming that the size of the eggs or hatchlings reflects the probability of survival of the offspring, we can get even closer to a full measure of fitness of the experimental animals. The size (cephalothorax length) of the young that emerged from the eggsacs was measured. The results show a clear decrease in the size of young from successive eggsacs (Fig. 2A); when analysing for diet effects it is therefore necessary to adjust for sac number. (It should be noted that such a size decline is not normally expected for the field situation, it is an experimental artefact (see above); the result shows, however, that fitness in the field may be sensitive to prolonged periods of monotonous prey availability, eg. as could result from insecticide spraying). There was also a significant parent-offspring correlation (Toft, in press), so the results were also adjusted for the size of the parents. What we see then is a significant effect of parent diet on the size of the young (Fig. 2b). The largest of the young are of the shortage-*Drosophila*+aphids and the shortage-*Drosophila*, followed by those of the rich *Drosophila* diet which were significantly smaller (Fisher's LSD test $p= 0.012$ and 0.051, respectively); finally young of the aphid treatment are significantly smaller still (all $p<0.001$). These size relationships strengthen the pattern we saw already for diet effects on reproductive success. In addition they will tend to improve the quality measure of the shortage *Drosophila* diet even more compared to the rich *Drosophila* diet.

The fact that both of the shortage diets produced larger young than the abundant diet, and that the nutritionally varied diet (aphid supplementation) made no further (significant) improvement, point to the interpretation of a life-history strategy that puts more resources into a smaller number of offspring when resources are scarce. However, the most food limited treatment obviously was the pure aphid diet, which showed the opposite response. I propose that this divergence can be explained by the spiders on the pure aphid diet being not only severely food limited, but at the same time under the influence of chemical defence substances (toxins?) of the aphids.

Using development and survival as quality criteria may produce conflicting results. Toft (in press) also compared the performance of first instar *E. atra* given aphids *R. padi* with that of unfed ones. Aphids greatly prolonged the survival of the spiders, but these never moulted, i.e. they were unable to develop. A similar experiment (S. Toft, unpubl.) with hatchlings of the wolf spider *P. prativaga* gave no improved survival compared with starving spiders when *R. padi* was given as food; however, other groups which were fed the aphids *Sitobion avenae* and *Metopolophium dirhodum*, respectively, showed increasingly longer survival. However, with none of these aphids did any of the spider young moult to the next instar. Thus, judged by survival, the three aphids were of different quality; judged by development they were all bad.

Sunderland et al. (1995) provide an example in which a diet, pollen+yeast, even appeared to have a negative effect on survival of the spider *Lepthyphantes tenuis* (Bl.) compared to starving spiders, though at the same time it promoted development (moulting) in some individuals.

Discussion

Prey quality is superficially a simple concept. However, as shown above, the same prey may be good by some criteria but bad by others. At the moment we know hardly anything about what properties may make a certain prey species good at some time to some predators, and which make them bad at other times to the same predators. The fact that different prey may be deficient in different nutritional properties, and that the predator consequently may improve the quality of its total diet by prey mixing, means that the quality of a prey type will vary depending on the immediate energetic and nutritional status of the individual. Still, very few cases of dietary self-selection are known from polyphagous predators (Greenstone 1979). The predators' needs as well as their ability to cope with chemically defended prey, are changing all the time as a function of their recent feeding history. Seen from the individual predators' point of view a prey species will therefore be of varying quality from one time to the next. The concept of prey quality should rather be considered in this more dynamic sense.

References

Bilde, T. & Toft, S. 1994. Prey preference and egg production of the carabid beetle *Agonum dorsale*. *Entomol. Exp. Appl.* 73: 151-56.

Chapman, R.F. & Sword, G.A. 1994. The relationship between plant acceptability and suitability for survival and development of the polyphagous grasshopper, *Schistocerca americana* (Orthoptera, Acrididae). *J. Insect Behav.* 7: 411-31.

Charnov, E.L. 1976. Optimal foraging: attack strategy of a mantid. *Amer. Natur.* 110: 141-51.

Cohen, A.C. 1995. Extra-oral digestion in predaceous terrestrial Arthropoda. *Annu. Rev. Entomol.* 40: 85-103.

Eisner, T. & Eisner, M. 1991. Unpalatability of the pyrrolizidine alkaloid-containing moth *Utetheisa ornatrix*, and its larva, to wolf spiders. *Psyche* (Cambr.) 98: 111-18.

Greenstone, M.H. 1979. Spider feeding behaviour optimises dietary essential amino acid composition. *Nature* 282: 501-3.

Hagen, K.S. 1987. Nutritional ecology of terrestrial insect predators. In: Slansky, F. & Rodriguez, J.G. (eds.) *Nutritional ecology of insects, mites, spiders, and related invertebrates*, pp. 533-77. John Wiley & Sons, New York.

Haukioja, E., Ruohomäki, K., Suomela, J. & Vuorisalo, T. 1991. Nutritional quality as a defense against herbivores. *Forest Ecology and Management* 39: 237-45.

Krebs, C.J. 1989. *Ecological methodology*. Harper & Row, New York.

Pulliam, H.R. 1975. Diet optimization with nutrient constraints. *Amer. Natur.* 109: 765-68.

Pyke, G.H. 1984. Optimal foraging theory: A critical review. *Ann. Rev. Ecol. Syst.* 15: 523-75.

Riechert, S.E. & Harp, J.M. 1987. Nutritional ecology of spiders. In: Slansky, F. & Rodriguez, J.G. (eds.) *Nutritional ecology of insects, mites, spiders, and related invertebrates*, pp. 645-72. John Wiley & Sons, New York.

Riechert, S.E. & Lockley, T. C. 1984. Spiders as biological control agents. *Ann. Rev. Entomol.* 29: 299-320.

Stephens, D.W. & Krebs, J.R. 1986. *Foraging theory*. Princeton Univ. Press, Princeton, USA.

Sunderland, K.D., Topping, C.J., Ellis, S., Long, S., Van de Laak, S. & Else, M. 1995. Reproduction and survival of linyphiid spiders in UK cereals. *Acta Jutlandica* (this volume).

Toft, S. (in press). Value of the aphid *Rhopalosiphum padi* as food for cereal spiders. *J. Appl. Ecol.*

Van Dijk, T.S. 1995. The influence of environmental factors and food on life cycle, ageing and survival of carabid beetles. *Acta Jutlandica* (this volume)

Waldbauer G.P. & Friedman, S. 1991. Self-selection of optimal diets by insects. *Annu. Rev. Entomol.* 36: 43-63.

Reproduction of beneficial predators and parasitoids in agroecosystems in relation to habitat quality and food availability

Sunderland, K.D.[1], *Bilde, T.*[2], *Den Nijs, L.J.M.F*[3], *Dinter, A.*[4], *Heimbach, U.*[5], *Lys, J.A.*[6], *Powell, W.*[7], *Toft, S*[2] *

[1]Entomology Department, Horticulture Research International, Littlehampton, West Sussex, BN17 6LP, UK
[2]Department of Zoology, University of Aarhus, Building 135, DK-8000, Aarhus C, Denmark
[3]Research Institute for Plant Protection, P.O. Box 9060, 6700 GW Wageningen, The Netherlands
[4]Institut für Pflanzenkrankheiten und Pflanzenschutz, Universität Hannover, Herrenhäuserstrasse 2, D-30419 Hannover, Germany
[5]Biologische Budesanstalt für Land- und Forstwirtschaft, Messeweg 11/12, D-38104 Braunschweig, Germany
[6]Zoological Institute, University of Berne, Baltzerstrasse 3, CH-3012 Berne, Switzerland
[7]Department of Entomology & Nematology, Rothamsted Experimental Station, Harpenden, Herts., AL5 2JQ, UK

* 2nd to 8th author in alphabetical order

Abstract

Manipulation of habitat quality and food availability are likely to be two of the main approaches for increasing natural enemy reproduction, which may contribute to maximising their abundance and effectiveness against pests. Factors affecting reproduction (in the broadest sense) that are reviewed here include larval ecology, quantity and quality of food, oviposition behaviour, egg mortality, cannibalism and the quality of eggs and hatchlings. Under each of these headings it is apparent that there is great variation, even between closely-related species of natural enemies, in optimal requirements for reproduction. Different stages in the life cycle of a single species

Arthropod natural enemies in arable land · II *Survival, reproduction and enhancement*
C.J.H. Booij & L.J.M.F. den Nijs (eds.). *Acta Jutlandica* vol. 71:2 1996, pp. 117-153
 ISBN 87 7288 672 2

requirements for reproduction. Different stages in the life cycle of a single species may also vary considerably in their requirements. Therefore, to maximise the abundance, diversity and quality of individuals in a natural enemy complex by increasing their reproduction it will be necessary to create conditions that offer a wide range of choice of food and physical conditions. Thus, a strategyof within- and between-field habitat diversification is appropriate for maximising natural enemy reproduction and promoting biological pest control.

Key words: Carabidae, Staphylinidae, Araneae, Coccinellidae, Syrphidae, Neuroptera, Heteroptera, Hymenoptera, Dermaptera, Acari, fecundity, food quantity, food quality, food preference, polyphagy, oviposition behaviour, egg mortality, egg quality, fertility, parasitism, cannibalism, larval ecology, distribution, host feeding, biological control, habitat diversification

Introduction

There is an acknowledged need to reduce pesticide inputs in agriculture (Pimentel 1995), but to maintain an acceptable level of pest control with less pesticide use, it will be necessary to find ways of maximising the impact of predators and parasitoids on pest populations. Measures are needed to decrease the mortality and increase the natality of natural enemies in agroecosystems, and to attract them into fields and to retain them during the cropping period. It is also important to find ways of concentrating their activities onto pest species (Finch, this volume). Manipulation of habitat quality and food availability are likely to be two of the main approaches available for increasing the numbers and effectiveness of natural enemies. The inter-relationships between mortality, natality and migration, and the relative importance of manipulating each of these processes, is beyond the scope of this review, which concentrates on reproduction. However, reproduction is considered here in a broad context, which extends to include all aspects from environmental influences on pre-adult stages through to consideration of factors affecting the quality of eggs and hatchlings (Fig. 1). Although empirical approaches to the augmentation of natural enemy reproduction are valuable in the short-term, the capacity to manipulate reproduction reliably, under any set of farming conditions, can only come from a detailed scientific understanding of the reproductive biology and ecology of the species concerned.

The aim of this review is to examine current knowledge concerning natural enemy reproduction as a precursor to recommending practical measures for augmenting reproduction in the field. Information is presented, in the main,

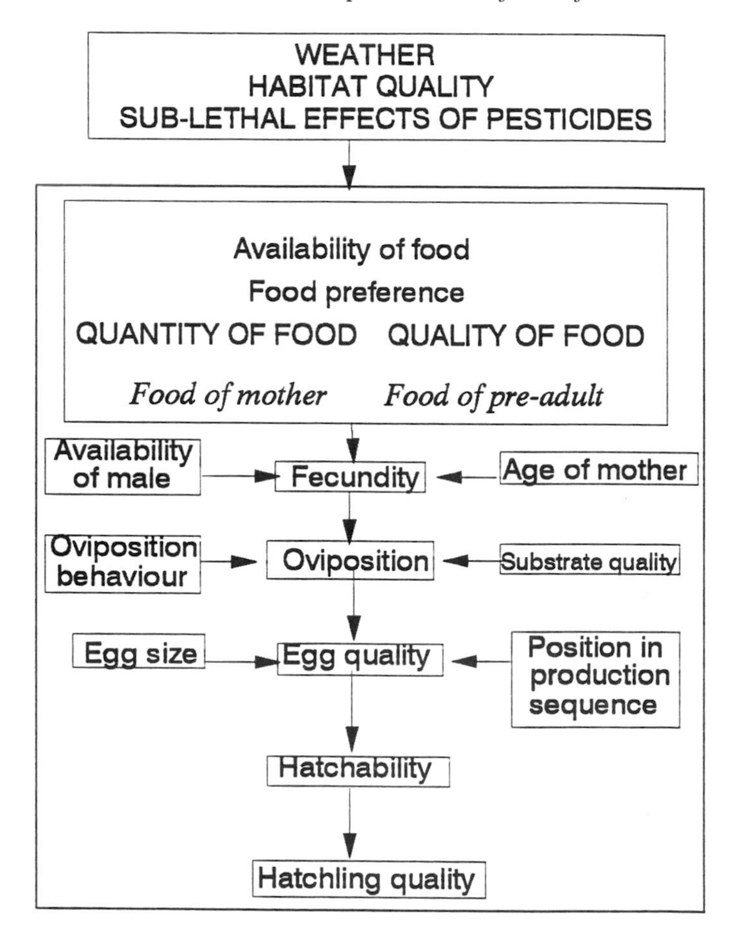

Fig. 1. Factors affecting reproduction.

for agricultural species of natural enemy, but occasionally also for closely-related species in non-agricultural habitats, to illustrate processes that could be occurring in agriculture. Pesticides can have diverse and subtle sublethal effects on all aspects of natural enemy reproduction. The considerable literature on this subject (see, for example, Jepson (1989) for a compilation of pesticide effects on natural enemies) is not included in this review, but it is worth noting that effects on reproduction constituted 30% of reports of deleterious side effects of pesticides on natural enemies in the extensive literature database of Theiling & Croft (1989). Attention is given first to the larval ecology of carabid and staphylinid beetles; this is a neglected area of research but is of importance here because the condition of the larva can affect the size and reproductive potential of the adult beetle (e.g. Nelemans 1988; Ernsting et al. 1992). Nutrition of the adult predator or parasitoid is a crucial factor; both quantitative and qualitative aspects are important and are discussed here. The significance of oviposition behaviour and the factors influencing egg mortality and the quality of eggs and hatchlings are also explored. Finally, consideration is given to the advantages and disadvantages of habitat diversification as a means of promoting natural enemy reproduction and reducing pest populations.

Factors affecting reproduction

Larval ecology of Carabidae and Staphylinidae

The larvae of some species of carabid forage regularly in the crop canopy (Sunderland 1992), many are active on the soil surface (Briggs 1965, Bauer 1982, Desender et al. 1985, Nelemans 1988, Wallin 1989, Sunderland 1992, Luff 1994) and others probably remain within the soil (Brunsting & Heessen 1984, Sunderland 1992). There is a similar vertical differentiation of foraging niches amongst staphylinid larvae inhabiting agricultural environments (Frank 1968, Heessen & Brunsting 1981, Kennedy et al. 1986, Sunderland 1992). Some surface-active carabid species are highly mobile and can travel tens of metres in the third instar (Nelemans 1988, Betz 1992) providing the opportunity for a degree of habitat selection. Such activity may be maintained even at low temperatures (Briggs 1965, Thiele 1977, Desender et al. 1985). Olfactory (Bauer 1982) or tactile (Spence & Sutcliffe 1982) cues are used to locate food, which, for agricultural carabids, can be seeds (Briggs 1965, Thiele 1977), aphids (Sunderland et al. 1987), Collembola (Bauer 1982, Nelemans 1988, Schelvis & Siepel 1988), earthworms (Thiele 1977, Schelvis & Siepel 1988), larvae of Diptera, larvae of elaterid, staphylinid and pterostichine beetles, mites and geophilomorph centipedes (Schelvis & Siepel 1988). Although there can be some dietary overlap between larvae and adults (Schelvis & Siepel 1988), carabid larvae sometimes have a more specialised diet than adults and in some species the larva is strictly carnivorous whilst the adult is omnivorous or phytophagous (Thiele 1977). Staphylinid larvae have been recorded to eat mildew (Dennis et al. 1991), mites, Diplopoda, Lumbricidae, Formicidae larvae, nematodes, insect eggs (Good & Giller 1988), aphids (Sunderland et al. 1987, Good & Giller 1988, Dennis et al. 1990), Collembola and the larvae of Diptera, Carabidae and Staphylinidae (Lipkow 1966, Good & Giller 1988) in the field.

Since larvae are often cryptic and difficult to study it is probable that their natural diets are very incompletely documented. Many species appear to have very limited food reserves when hatching from the egg and finding a first meal quickly can be critical for survival (Heessen & Brunsting 1981, Heimbach, unpubl.). At low temperatures, however, some carabid larvae can survive without food for more than a month (Luff 1994). Larval development is affected by the quantity and quality of food consumed (Theiss & Heimbach 1993, 1994b); for example larvae of the carabid *Nebria brevicollis* must double in weight before moulting (Nelemans 1987). A single prey type, such as pupae of Diptera, can support successful development by some carabid species (Heimbach unpublished), but many single-prey diets are inadequate and result in failure to moult or a high mortality; this was the case for larvae of the carabid *Pterostichus cupreus* (= *Poecilus cupreus*) reared on Collembola or aphids (Theiss & Heimbach 1993) and larvae of the staphylinid genus *Tachyporus* reared on pollen or yeast (Lipkow 1966).

In addition to inadequate nutrition, larval mortality can result from natural enemies and adverse physical conditions. Few generalisations can be made about the abiotic conditions favoured by larvae, except that very dry microclimates are detrimental. Some species suffer a high mortality in response to waterlogging of the soil (Theiss & Heimbach 1994a), low temperature (Ernsting & Huyer 1984) or high temperature (Metge & Heimbach 1994). The staphylinid genus *Tachinus* requires a relative humidity of 100% for development, but a closely related genus, *Tachyporus,* can tolerate much lower humidities (Lipkow 1966). Similar variations between species in optimal physical conditions apply also to sympatric species of carabid in non-agricultural habitats (Van Dijk, this volume). During laboratory rearing of carabids and staphylinids Heimbach (unpublished) has observed that larvae often require to dig holes for moulting and that success in doing this varies with soil type. Destruction of moulting holes (as might occur during some farming operations) usually results in deformity of the larva or mortality, but there are no quantitative data from the field. Little is known either about the impact of natural enemies on larvae, but Nelemans et al. (1989) estimated *c.* 20% mortality of *N. brevicollis* larvae due to attack by entomogenous fungi and parasitoids. Larval cannibalism, however, is considered to be common amongst carabids (Thiele 1977, Brunsting & Heessen 1984, Nelemans 1988) and has also been recorded in Staphylinidae (Heessen & Brunsting 1981). Larvae of the carabid *Pterostichus oblongopunctatus* are cannibalistic even at low density (Heessen & Brunsting 1981) and cannibalism is considered to be the main factor regulating population density in this species (Brunsting & Heessen 1983, Brunsting et al. 1986).

Fecundity

The definition of *fecundity* used in this paper is the total number of eggs produced by an individual female during her reproductive life, whereas *egg production* is the number of eggs produced over a given period of time and *egg complement* refers to the number of eggs in the ovaries at any given time. There is a correlation between egg complement and fecundity in some species of carabid, but not in others (Van Dijk 1979); the relationship has not been investigated for the majority of species of predator and parasitoid. An indication of the range of fecundities typical of agricultural species of predator, based on a limited number of publications, is 12-223 for Linyphiidae (De Keer & Maelfait 1987, 1988, Sunderland et al., this volume), 17-230 for Staphylinidae (Lipkow 1968, Eghtedar 1970, Heessen 1980, Kennedy et al. 1986, Metge & Heimbach 1994), 5-900 for Carabidae (Briggs 1965, Grüm 1973, Hůrka 1975, Thiele 1977, Heessen 1980, Luff 1982, 1987, Desender et al. 1985, Nelemans 1987, Heimbach 1989, Nelemans et al. 1989, Weseloh 1993), and 160-3765 for Coccinellidae (Hämäläinen & Markkula 1972, Rogers et al. 1972a, Ruzicka et al. 1981, Hemptinne et al. 1988). Although this may seem to suggest that Coccinellidae have the greatest capacity for increase, such a conclusion can only be based on *fertility* data (i.e. the total number of viable eggs produced by a female during her reproductive life), and it is possible that many Coccinellidae produce a high proportion of infertile eggs as food for neonate larvae (see section on Cannibalism).

Effects of food quantity on egg production

Agricultural carabids produce more eggs in response to higher rates of food consumption (Heessen 1980, Ernsting & Huyer 1984, Wallin et al. 1992, Van Dijk, this volume) as do carabids in non-agricultural habitats (Baars & Van Dijk 1984, Sota 1984, 1985, Juliano 1986, Van Dijk 1994). Eggs that have already been formed can be resorbed (oosorption) in response to a food shortage (Mols 1988) and the response time can be as little as two days in some species (Van Dijk 1982, 1986). Size of carabid adult, feeding rates of adults taken from the field and various other considerations suggest that carabid populations probably experience a more or less continuous shortage of food in the field (Baars & Van Dijk 1984, Sota 1985, Nelemans 1987, Bilde & Toft, this volume, Van Dijk, this volume) and this may be a general experience for predators (White 1978). Egg complement or egg production by carabids can vary significantly between crops (eg *Pterostichus cupreus* produced more eggs in sugar beet than winter wheat; Den Nijs, unpubl.) or between sites (e.g. great variation in egg complement of

P. oblongopunctatus between forest stands; Szyszko, this volume), but the causes of such variation remain to be established. Herbicide applications to the edges of cereal crops reduced the abundance of phytophages and reduced the quantity of food in carabid guts and their egg complements (Chiverton & Sotherton 1991). Female *P. cupreus* from cereals near weed strips had fuller crops and contained more eggs than those from control areas (Zangger et al. 1994). The above considerations suggest that there is considerable potential for increasing recruitment rates of carabids by habitat management to increase food availability.

Egg production by spiders is greater when the food supply is increased and this has been recorded for Linyphiidae (De Keer & Maelfait 1987, 1988, Alderweireldt & Lissens 1988) and Lycosidae (Kessler 1971, Suzuki & Kiritani 1974) from agricultural habitats and for a range of families from non-agricultural habitats (Turnbull 1962, Wise 1975, 1979, Kessler 1973, Van Wingerden 1978, Fritz & Morse 1985, Miyashita 1992, Spiller 1992). As for Carabidae, authors report indirect evidence of food shortage commonly experienced by spiders in the field (Miyashita 1968, Van Wingerden 1977, Fritz & Morse 1985, De Keer & Maelfait 1987), and, in some cases, consistent differences between sites have been observed over several years (Miyashita 1992). Some species of the lycosid genus *Pardosa* respond to moderate food shortage by reducing egg production immediately but other congeneric species maintain a consistent level of egg production, except under conditions of extreme food shortage (Kessler 1971). Females of the linyphiid *Linyphia triangularis* sometimes became sterile as a result of inadequate food availability during the pre-adult stages (Turnbull 1962). Other groups of predator in which egg production can be affected by the quantity of food consumed include Coccinellidae (Dixon 1959, Ghanim et al. 1984, Dixon & Guo 1993), Heteroptera (Anderson 1962, O'Neil & Wiedenmann 1990, DeClercq & Degheele 1992) and Diptera (Morris & Cloutier 1987). In the Coccinellidae potential egg production is affected by the food supply available not only to the adult but also to the larva because adult size and number of ovarioles is affected by larval nutrition (Dixon & Guo 1993).

Most adult female parasitoids feed to fuel their host finding and reproductive activities. Pro-ovigenic species, which complete development of their eggs soon after emergence, usually only require a source of carbohydrates and feed mainly on nectar and the honeydew produced by homopterans such as aphids and scale insects (Stary 1970, Gauld & Bolton, 1988). Synovigenic species, which develop their eggs successively over a long adult life span, also require a protein source and this is often obtained directly from their hosts. However, regardless of oogenesis strategy, the quantity and quality of food available to adult females can affect their realised fecundity. Lack of food can result in resorption of eggs or prevent the development of eggs in synovigenic species (King, 1963, King & Richards, 1968, Jervis & Kidd, 1986, van Lenteren et al., 1987).

Effects of food quality on egg production by predators

Important aspects of food quality include chemical defences of the prey (feeding deterrents and toxins), energy content and nutritive value. Food quality must, however, be considered as a *dynamic* concept because it can vary according to the previous nutritional history of the predator (Toft, this volume). For example, a prey rich in a particular amino acid would be of high quality to a predator deficient in that amino acid but not necessarily of high quality to another individual predator of the same species that was not so deficient.

Pyrrolizidine alkaloids sequestered from the food plant protect eggs, larvae and adults of the moth *Utetheisa ornatrix* from attack by coccinellid beetles and argiopid and lycosid spiders. Laboratory-reared alkaloid-free moths were readily accepted by these predators but mealworms (normally a high quality food supporting reproduction) treated with alkaloids were rejected (Eisner & Eisner 1991). Sometimes, however, predators can adapt behaviourally to the presence of toxins in prey. For example, the spider *Cupiennis salei* adapted to eating crickets experimentally poisoned with KCN by ingesting mainly the lipid component of the prey and thus avoiding the highest concentrations of the water-soluble KCN (Nentwig 1985). Fecundity of the linyphiid spider *Erigone atra* given the cereal aphid *Rhopalosiphum padi* was very low, and presence of a toxin or deterrent was suspected because the consumption rate of this prey was minimal (Toft 1995). Egg production by the carabids *Pterostichus cupreus* (Wallin et al. 1992) and *Agonum dorsale* (Bilde & Toft 1994) on a diet of this aphid was much less than on other foods. *Coccinella septempunctata* laid 66% fewer eggs on a diet of *R. padi* than on a diet of the cereal aphid *Sitobion avenae* (= *Macrosiphum avenae*) (Ghanim et al. 1984). A reduced preference of *A. dorsale* for fruit flies coated with *R. padi* homogenate, compared with control flies (Bilde & Toft 1994), indicates that this aphid may contain a feeding deterrent that is active against a wide range of predators. On the other hand, various species of aphid, thought to contain toxins, have been found to be detrimental to egg production in some species of Coccinellidae whilst being perfectly adequate to support reproduction by other coccinellid species (Hodek 1957, Blackman 1967), suggesting that some toxins are quite specific. Dietary constituents can also affect predator reproduction in other ways; for example some foods have been found to initiate reproductive diapause in heteropteran bugs of the genus *Anthocoris* (Anderson 1962).

Some foods (*essential* foods) are adequate to support reproduction whilst others (*alternative* foods) prolong survival but do not, by themselves, support reproduction (Mills 1981). Various reported alternative foods include plant foods (Sota 1984, Weseloh 1993), Collembola and dipteran larvae (Nelemans 1987) for some Carabidae, yeast (Lipkow 1966) for some Staphylinidae, pollen, yeast, aphids, mites (Sunderland et al. this volume) and Diptera (Miyashita 1968) for some spiders. The predatory mites *Gamasellodes vermivorax* and *Macrocheles* sp. required nutrients from nematodes for reproduction and could not produce eggs on an arthropod-only diet (Walter et al. 1987).

Essential foods vary greatly in the level of egg production that they support and this has been documented for carabids (Van Dijk 1986, Weseloh 1993, Van Dijk this volume), coccinellids (Blackman 1967, Hämäläinen & Markkula 1972, Rogers et al. 1972a, Ghanim et al. 1984, Hodek 1993), chrysopids (Sengonca et al. 1987), predatory Diptera (Zöllner & Poehling 1994), heteropteran bugs (Kiman & Yeargan 1985, Ruth & Dwumfour 1989), spiders (Sunderland et al. this volume, Toft, this volume) and predatory mites (McMurtry & Scriven 1964, Zhao & McMurtry 1990, Ouyang et al. 1992). Foods that are suitable for development of pre-adult stages are sometimes (Ruberson et al. 1991) but not always also suitable for egg production by the adult; examples are to be found in the Coccinellidae (Blackman 1967) and Heteroptera (Anderson 1962). The quality of herbivorous prey for supporting predator reproduction can be affected by the quality of the herbivore's host plant and whether the herbivore is healthy or diseased. In general, aphids feeding on Leguminosae are often of poor nutritional quality for predators (Zöllner & Poehling 1994). Fecundity of the heteropteran bug *Nabis roseipennis* was lower when feeding on soybean looper caterpillars infected with a nuclear polyhedrosis virus compared with its fecundity on a diet of healthy caterpillars (Ruberson et al. 1991).

The majority of predator species in agroecosystems are polyphagous and it is important to know whether such polyphagy is an accidental outcome of temporal shifts in prey availability or whether it is a physiological necessity for successful growth, development and reproduction. It is possible to rear some species of predator successfully over many generations in the laboratory on single-species diets without any apparent loss of vigour or productivity and a few examples are reported amongst the Carabidae (Thiele 1977, Heimbach, unpubl.) and Acari (McMurtry & Scriven 1964). Other authors report that predators reached maturity and reproduced on single-species diets, but egg viability and fitness of the offspring was not investigated (Turnbull 1962, Lipkow 1966, Holmberg & Turnbull 1982). In the majority of cases a mixed diet resulted in greater egg production than a single-species diet; instances are to be found in the Carabidae (Wallin et al. 1992, Bilde & Toft 1994), Heteroptera (Bush et al. 1993), Araneae (Miyashita 1968, Suzuki & Kiritani 1974, Greenstone 1979, Holmberg & Turnbull 1982, Thang et al. 1990, Uetz et al. 1992, Toft, this volume) and Acari (McMurtry & Scriven 1964, Zhao & McMurtry 1990). It seems likely that, in general, more than one prey type is needed by predators to maximise fitness. High quality food is limited in the field and polyphagy, with associated mechanisms of prey switching (Murdoch 1969) and dietary self-selection (Waldbauer & Friedman 1991) might be a strategy for securing the calories and diversity of nutrients needed for growth, development and reproduction (Van Dijk, this volume). Some *Pardosa* spiders are thought to be able to select prey species to optimise the proportions of essential amino acids in their diet (Greenstone 1979, but see also criticism by Humphreys 1980). In laboratory trials *Pardosa vancouveri* preferred the prey that promoted maximum fitness (inluding fecundity) (Holmberg & Turnbull 1982) but the predatory mite *Amblyseius hibisci* showed no preference between mites and pollen, even though reproduction was significantly better on a pollen diet (McMurtry & Scriven 1964). Far more examples will be needed before generalisations are possible about the extent to which predators have the capacity to control dietary intake to maximise fitness.

Effects of food quality on egg production by parasitoids

Food quality, in both the larval and adult stages, may affect parasitoid fecundity and longevity. For parasitoids reared on artificial diets, omission of amino acids, inorganic salts or vitamins reduced fecundity (Jervis & Kidd 1986). As in Carabidae (above) many parasitoids can respond to inadequate nutrition by oosorption. *Aphytis melinus* (Aphelinidae) females resorbed eggs when they were fed on sucrose only, but addition of yeast to the diet greatly reduced the problem (Heimpel & Rosenheim 1995). Similarly, poor nutrition resulted in oosorption by the ichneumonids *Itoplectis conquisitor* and *Scambus buolianae*, and the greatest fecundity occurred when females were given a mixed diet of carbohydrate and proteins (Leius 1961a, 1961b). When hosts are lacking some species of parasitoid display a rate of oosorption that equals the rate of oogenesis; this means that a fairly constant number of eggs reach maturity each day and the female is able to exercise restraint in oviposition and choice of hosts. In this way oosorption can play an important role in host selection and the efficiency of parasitization (Flanders 1942). In some species, partially resorbed eggs can be laid into hosts; the probability of fertilization of these eggs may be different from that of un-resorbed eggs with concomitant repercussions for parasitoid sex-ratios (Jervis & Kidd 1986). Protein diets did not increase the fecundity or adult longevity of the lepidopteran egg parasitoid *Trichogramma* but honey and cotton plant nectar did, compared with unfed wasps (Ashley & Gonzalez 1974).

Nectar and pollen from flowering plants are an important food source for many adult parasitoids (Van Emden 1962, Leius 1963, Jervis et al. 1993) and the presence of flowers in the vicinity of crops can increase levels of parasitism in populations of pests on those crops (Leius 1967, Topham & Beardsley 1975, Foster & Ruesink 1984, Powell 1986, Jervis et al. 1992). Access to a range of flowering plants significantly increased the fecundity of the eulophid parasitoid *Hyssopus thymus*, an important natural enemy of the European pine shoot moth (Syme, 1975). After monitoring the parasitic Hymenoptera visiting a range of flowering plants in England, Jervis et al. (1993) suggested the use of umbelliferous plants as an adult parasitoid food source in pest management strategies.

Adult nutrition also affects longevity (Jervis & Kidd 1986) and this can influence fecundity in both synovigenic and pro-ovigenic parasitoids. Obviously, in synovigenic species reduced longevity means fewer eggs can be developed, but in pro-ovigenic species enough time is also needed to allow the full egg load to be utilised. The oviposition rate of ten eggs per day recorded for the braconid *Orgilus obscurator* required a longevity of three weeks for the full egg load of 160-200 to be laid, but females only lived that long if they were allowed to feed on certain flower species (Syme 1977). *Trichogramma pretiosum* females lived three times as long when fed on a honey/nectar mixture, regardless of the presence or absence of hosts (Treacy et al. 1987). Carbohydrates are also needed to fuel the foraging activity of female parasitoids, and to be useful in pest management parasitoids need to be efficient at low host densities when host location requires more energy consumption. Therefore the closer the food source is to the host population the more effective the parasitoid is likely to be, since energy is not wasted in flying long distances between host and food locations. Plants with extrafloral nectaries are a useful food source (Jervis et al. 1993) and parasitism of *Heliothis zea* eggs by *T. pretiosum* was reduced significantly in plots of a nectariless cotton cultivar compared with that in nectaried cultivar plots (Treacy et al. 1987).

Many parasitoid species preferentially visit and feed from specific flowers (Leius 1960, Syme 1975, Jervis et al. 1993). Also, a mixture of food types may be necessary for maximum fecundity to be achieved. In laboratory studies of *S. buolianae* the greatest fecundity was obtained when females were fed host larvae plus pollen in a sucrose solution (Leius 1961b).

Fecundity can be affected by the nutritional quality of the host in which the female parasitoid developed, which in turn is affected by the diet or host plant of the host (Orr et al. 1985, Van Emden & Wratten 1991). The fecundity of the egg parasitoid *Telenomus chloropus* (Scelionidae) was reduced by a half when it was reared on eggs produced by the stink bug *Nezara viridula* feeding on a resistant soybean cultivar compared with that of parasitoids reared on eggs produced by hosts feeding on a susceptible cultivar (Orr et al. 1985). In contrast, the fecundity of the tachinid parasitoid *Lydella thompsoni* was not affected when it was reared on corn borer larvae (*Ostrinia nubilalis*, Lepidoptera) that had been fed on maize plants of different genotypes (Sandlan et al. 1983). Presence of entomopathogens in hosts during parasitoid development can also affect the fecundity of emerging females. The fecundities of several parasitoids of lepidopteran larvae are significantly reduced if they are reared on hosts infected with *Bacillus thuringiensis* (Hamed 1979, Temerak 1980, Salama et al. 1982, Salama et al. 1991).

The honeydew produced by aphids, scale insects and other Homoptera is an important food source for some adult parasitoids, which may feed on it as it is excreted or after it has been deposited on leaf surfaces (Jervis et al. 1992). Braconid parasitoids of aphids not only feed on the honeydew excreted by their hosts but use it as a host location cue (Budenberg 1990). *Edovum putteri*, a eulophid egg parasitoid of the Colorado potato beetle (*Leptinotarsa decemlineata*) also feeds on homopteran honeydew, and absence of aphids in the crop may hinder the early-season establishment of the parasitoid (Idoine & Ferro 1988). However, honeydew is not a useful food for all parasitoids. When honeydew from two aphid species was fed to *S. buolianae*, a moth parasitoid, it reduced the fecundity of females and the longevity of both sexes (Leius 1961b).

Many synovigenic parasitoids obtain the proteins necessary for continued egg development by feeding on the haemolymph or body tissues of their hosts, often but not always using different individuals for host feeding and oviposition (Edwards 1954, Leius 1962, 1967, Bartlett 1964, Jervis & Kidd 1986, van Lenteren et al. 1987, Jervis et al. 1992). In some Pteromalidae host feeding is an obligatory precursor to oviposition (Flanders 1942). Host feeding has been recorded in 17 families of parasitic Hymenoptera (Jervis & Kidd 1986) as well as in the dipteran family Tachinidae (Nettles 1987). When the aphelinid *Aphytis melinus* uses the same oleander scale (*Aspidiotus nerii*, Diaspididae) for both oviposition and host feeding, the immature parasitoid fails to complete development (Heimpel & Rosenheim 1995). The decision to use a particular host for either oviposition or host feeding is influenced by the current egg load of the parasitoid and may also be influenced by host size, smaller hosts being more frequently used for host feeding (Heimpel & Rosenheim 1995). The lifetime fecundity of the dryinid *Dicondylus indianus* attacking the rice brown planthopper (*Nilaparvata lugens*, Delphacidae) had a strong positive relationship to the number of hosts fed upon (Sahragard et al. 1991). The rate of host feeding by *D. indianus* was greater at higher host densities, leading to increased fecundity with greater host availability (Sahragard et al. 1991). Low host densities will result in more energy being expended in host location and hence less available resources for egg maturation. Jervis & Kidd (1986) extensively review the subject of host feeding by parasitoids.

Other factors affecting egg production

Egg production by predators and parasitoids may be affected by abiotic conditions, including photoperiod (Ruzicka et al. 1981) and temperature (Hämäläinen et al. 1975). Fecundity can be greater at the lower end of the range of temperatures that the predator experiences in the field (Wright & Laing 1978, Ernsting & Huyer 1984, Palanichamy 1985, Sunderland et al. this volume) or the reverse (Morris & Cloutier 1987, Braman & Yeargan 1988, De Keer & Maelfait 1987, 1988) depending on species. In some cases fecundity is less at high temperatures because the reproductive lifespan of the female is reduced (Palanichamy 1985). Some carabids require fluctuating temperatures for egg production (Heessen 1980) or egg production is greater under fluctuating than constant temperatures (Ernsting & Huyer 1984).

In the majority of species that have been investigated egg production is positively correlated with the size or biomass of the mother, and this is true for carabids (Ernsting & Huyer 1984, Juliano 1985, Zangger 1994, Heimbach unpublished, Szyszko, this volume), coccinellids (Stewart et al. 1991a), chrysopids (Canard & Duelli 1984), spiders (Hackman 1957, Kessler 1971, Fritz & Morse 1985), heteropterans (Evans 1982, O'Neil 1992) and parasitoids (Godfray 1994). The relationship with the age of the mother is, however more variable (NB to simplify presentation in this review, maternal age of spiders will be equated with sac number, because later eggsacs in the production sequence will often have been produced by older mothers). Depending on species of predator or parasitoid, younger females may produce more (Van Dijk 1979, Jervis & Kidd 1986, De Keer & Maelfait 1988, Willey & Adler 1989, Dinter & Poehling 1995) or fewer (Edgar 1971b, Kessler 1973, Van Dijk 1979) eggs than older females, or there may be no correlation between egg production and maternal age (Sota 1984, Sunderland et al. this volume).

Repeated mating by the same pair of individual carabid beetles has been observed (Briggs 1965) and continuous presence of a male has been found to increase egg production in a number of carabid species (Van Dijk 1973, Nelemans 1987). At low mating frequencies the carabid *Pterostichus coerulescens* (= *P. versicolor*) will resorb its eggs even when the food supply is adequate (Mols 1988). Continuous presence of a male also increases egg production in some species of staphylinids (Heimbach, unpubl.), predatory Diptera (Morris & Cloutier 1987) and earwigs (Shepard et al. 1973), but production of a second brood of the earwig *Forficula auricularia* did not require further mating (Good 1982). Some species of parasitoid cause sterilisation of their heteropteran (O'Neil 1992) and coccinellid hosts. The coccinellid *Coleomegilla maculata* is sterilised by *Perilitus coccinellae* (Wright & Laing 1978) but *Coccinella 7-punctata* can sometimes resume reproduction after emergence of *P. coccinellae* (Ruzicka et al. 1981).

Some species of predatory mite are sensitive to the density of conspecifics and reduce egg production at higher densities, but others do not respond in this way (Yao & Chant 1990). Fecundity of carabids can be reduced at high population density (Brunsting & Heessen 1984, Heessen 1980), but if density is too low fecundity may also fall because of a reduced mating frequency (Mols 1988) suggesting that there is an optimal density for the maximisation of fecundity.

Amongst parasitoids of the families Braconidae and Tachinidae, species attacking later host instars tend to have a lower fecundity than those attacking earlier instars; this may have evolved because the probability of host mortality before parasitoid emergence is less in the former case (Braman & Yeargan 1991).

Van Dijk (1982) reported that 67% of the variance in egg production of some carabids, under conditions of optimal food at constant temperature in the laboratory, was attributable to differences between individuals of the same species, which suggests that intraspecific genetic variation could be a very significant factor affecting the egg production capacity of predators.

Oviposition

Oviposition behaviour is often influenced by physical conditions such as temperature and soil moisture. Eggsacs of the linyphiid genus *Oedothorax* are rarely constructed at temperatures below 10°c (Alderweireldt & De Keer 1988) and 13°c is the lower temperature threshold for mating and oviposition by the coccinellid *Propylea 14-punctata* (Hemptinne et al. 1988). Some carabids (Van Dijk & Den Boer 1992) and staphylinids (Lipkow 1968) prefer to oviposit in moist rather than dry soil. In some lycosid species females carrying eggsacs look for exposed sites on sunny days to warm their eggsacs (De Keer et al. 1989). The predatory mite *Pergamasus crassipes* usually oviposits in the cool of the night (Bhattacharyya 1962) as does the lycosid *Pardosa lugubris* (= *Pardosa chelata*) (Vlijm & Richter 1966). In this species the pre-mating courtship display incorporates palpal drumming and a substrate is favoured which makes perception of the drumming possible (Hallander 1970).

A feeding stimulus may be required before oviposition (Zangger 1994), or the oviposition rate may be increased after feeding in various species of Carabidae (Van Dijk 1986, Ernsting et al. 1992) and predatory Diptera (Morris & Cloutier 1987). Some carabids (Nelemans 1987) and linyphiids (De Keer & Maelfait 1988, Sunderland, unpubl.) continue to oviposit when the eggs are unfertilized (Nelemans 1987), but mating can stimulate oviposition in other carabids (Van Dijk 1973). In *Coenosia tigrina* (a dipteran predator of dipterous pests) accessory secretions of the male supplement female nutrition, accelerate egg maturation and stimulate oviposition (Morris & Cloutier 1987). Mating is essential for oviposition by *Pergamasus crassipes* (Bhattacharyya 1962).

The nature of the substrate can influence choice of oviposition site because there are differences between species in the extent to which eggs are protected. Some species of carabid prefer a humus soil (Heimbach, unpubl.), some will not oviposit unless clay is available (Thiele 1977) and others show little preference (Heimbach, unpubl.). Carabid eggs are laid singly or in groups (Luff 1987) on or near the soil surface (Briggs 1965, Van Dijk & Den Boer 1992) or in cavities or mud cells within the soil (Sota 1985, Van Dijk & Den Boer 1992) and the eggs are sometimes coated with the substrate (Luff 1987). These measures are thought to protect the eggs from desiccation and attack by entomogenous fungi (Brandmayr & Brandmayr 1979). There is a similar range of behaviour in the Staphylinidae, from deposition of single unprotected eggs (Frank 1968, Lipkow 1968) to coating eggs with clay and placing them in crevices or specially constructed chambers (Lipkow 1968, Kennedy et al. 1986). Parental care of the eggs may extend to the guarding of eggsacs by spiders (Toft 1979, Willey & Adler 1989, Van Baarlen et al. 1994) and egg cleaning, temperature control and protection of hatchling nymphs by earwigs (Good 1982). Predators, such as syrphid flies, coccinellid beetles and heteropteran bugs, which oviposit into and onto living plant material, are sensitive to its surface texture and oviposition rates may vary with the type of plant available (Chandler 1968c, Braman & Yeargan 1988) and also with substrate colour, degree of curvature and amount of shading (Iperti & Prudent 1986, Iperti & Quilici 1986). The coccinellid *Adalia bipunctata* is stimulated to oviposit by olfactory cues from Juniper, Cypress and Fennel (the reason is unknown) (Boldyrev et al. 1969, Iperti & Prudent 1986) and eggs of other agricultural species have been found clustered on human litter in the field (Banks 1956, Sunderland, unpubl.). There is much to be learnt about the stimuli eliciting oviposition and such knowledge might enable the distribution of predator eggs in the field to be manipulated (Boldyrev et al. 1969).

The distribution of eggs can be influenced by the distribution of conspecifics and prey, and by plant and habitat structure. Amongst the Coccinellidae are to be found aphidophagous aphidozetic species that prefer to oviposit on young leaves harbouring young aphid colonies (Hemptinne et al. 1993), or on leaves with aphid colonies large enough to promote survival of neonate larvae (Wratten 1973) but there are also phytozetic aphidophagous species that oviposit on plants with no aphids present (Banks 1956, Kawai 1978). The oviposition of some coccinellid species is inhibited by the presence of conspecific larvae (Hemptinne et al. 1992) or adults (Hemptinne et al. 1993), but other species are not inhibited by conspecific eggs or larvae (Hemptinne et al. 1993). Similarly, there are phytozetic and aphidozetic Syrphidae (Chandler 1968a) and, amongst the latter group, species vary in the size or colour morph of aphid colony that stimulates oviposition (Chandler 1968b). The distance from an aphid colony at which eggs are deposited can also alter with the age of ovipositing female (Chandler 1967). Some syrphids are deterred from ovipositing by the presence of conspecific larvae (Krause & Poehling, this volume) but others are undeterred by conspecific eggs, larvae or the expelled gut contents of larvae (Chandler 1968a). *Episyrphus balteatus* oviposits near small aphid colonies (Kan 1989) and other species avoid ovipositing near colonies that contain alatae (Krause & Poehling, this volume) reducing the probability that the aphids will disperse before the syrphid eggs hatch. Species vary in their preferred oviposition locations on the plant, including the undersides of leaves (Chandler 1968c) and the proximity of leaf edges (Chandler 1968a). However, the great mobility of syrphid adults also endows them with the capacity to choose oviposition sites on the habitat scale. Many choose to oviposit in fields near to hedges (Krause & Poehling, this volume) and numerical responses to aphid density can be made over several thousand hectares of farmland (Chambers 1991).

Unlike predators, parasitoids have the added problem of finding hosts before they can oviposit and in some species it is this factor, rather than egg production, that limits reproduction (Wright & Laing 1978). Because of this problem, most parasitoids with adequate egg supplies probably oviposit in nearly all hosts they encounter even where the survival probability of the progeny is low (Godfray 1994).

Egg mortality

Mortality of eggs, due to causes other than infertility and cannibalism (discussed in later sections), can have a significant effect on recruitment rates of predator and parasitoid populations, but few data are available on the biotic, and especially the abiotic, causes of mortality. Carabid eggs suffer mortalities of 50-80% under field conditions (Heessen 1981, Nelemans et al. 1989, Van Dijk & Den Boer 1992), but this can be reduced to less than 10% if predators are eliminated experimentally (Heessen 1981). The identity of these egg predators is not known, but nematodes are suspected (Heessen 1981, Van Dijk & Den Boer 1992). In non-agricultural habitats spider egg mortality ranged from 10-60% due to parasitoids ovipositing into their eggsacs (Edgar 1971a, Van Wingerden 1973, Kessler & Fokkinga 1973, Rollard 1990). Eggs can also be eaten by ants, orthopterans, cantharid beetles and spiders (Willey & Adler 1989) or be killed by low temperatures (Steigen 1975). In some species of linyphiid nearly 30% of eggsacs are attacked by parasitoids in winter wheat (Van Baarlen et al. 1994) and mortality of eggs and hatchlings of the linyphiid *Lepthyphantes tenuis* has been estimated at 98% in winter wheat (Topping & Sunderland, this volume). Egg masses of a wide range of predatory pentatomid heteropterans are attacked by polyphagous scelionid parasitoids (Braman & Yeargan 1989). Although coccinellids often eat conspecific eggs (see section on Cannibalism below) they are reluctant to eat conspecific eggs experimentally painted with water extracts of eggs of other species of Coccinellidae, suggesting chemical protection against hyperpredation, perhaps by the alkaloids coccinelline and adaline (Agarwala & Dixon 1992). Alkaloids in coccinellid eggs deter ant predation and the odour of defensive chemicals probably deters predators when eggs are clustered because a threshold concentration is exceeded (Agarwala & Dixon 1993b).

Both solitary and gregarious parasitoids have been observed to practice conspecific ovicide (often using the ovipositor), and interspecific ovicide using toxins has also been recorded (e.g. *Ephedrus* spp. inject a toxin into the aphid host at oviposition which kills *Aphidius* spp. eggs but has no effect on *Ephedrus* spp. eggs) (Godfray 1994).

Cannibalism

Cannibalism, including cannibalism of eggs, provides all the nutrients needed by predators; for example, larvae of some syrphid and coccinellid species can be reared to normal adults on an exclusive diet of conspecific eggs (Chandler 1969, Ng 1986a). Conspecific eggs are a better food than aphids for larval growth and survival of some Coccinellidae (Agarwala & Dixon 1992, 1993a). Cannibalism can be a valuable tactic to allow a proportion of a population to survive and develop during periods of food shortage. It occurs extensively amongst predators and has been reported in the Carabidae (Thiele 1977, Heessen & Brunsting 1981, Brunsting & Heessen 1983, Brunsting & Heessen 1984,

Brunsting et al. 1986, Nelemans 1988), Staphylinidae (Heessen 1980, Heessen & Brunsting 1981, Heimbach, unpubl.), Coccinellidae (Banks 1955, Dixon 1959, Pienkowski 1965, Witter 1969, Rogers et al. 1972b, Kawai 1978, Ruzicka et al. 1981, Mills 1982, Ng 1986a,b, Agarwala & Dixon 1992, 1993a,b), Diptera (Chandler 1969, Morris & Cloutier 1987), Heteroptera (DeClercq & Degheele 1992), Dermaptera (Shepard et al. 1973, Good 1982), Araneae (Turnbull 1962, Hallander 1970, Valerio 1974, Schaefer 1977, Van Wingerden 1977, Kenmore et al. 1984, Nentwig 1987, Willey & Adler 1989, Thang et al. 1990, Sunderland et al. 1994) and Acari (Bhattacharyya 1962). In some species cannibalism only occurs during food shortage (Heessen 1980) but in others it is recorded even in the presence of abundant alternative food (Banks 1955, Bhattacharyya 1962, Hallander 1970, Morris & Cloutier 1987). It can account for a significant proportion of food intake in the field; for example 30% of the prey of the lycosid *Pardosa pullata* were conspecifics (Hallander 1970) and 70% of coccinellid larvae were observed feeding on eggs of their own species (Banks 1955). Van Wingerden (1977) estimated a mortality rate of 8% per week due to cannibalism by the linyphiid *Erigone arctica* and cannibalism reduced larval densities of the coccinellid *Adalia bipunctata* by 8-30% in different years (Mills 1982). Intense multi-stage cannibalism in the coccinellid *Propylea 14-punctata* resulted in only 9.4% of eggs reaching the adult stage (Rogers et al. 1972b). Sometimes females eat their own eggs (Good 1982, Canard & Duelli 1984, DecClercq & Degheele 1992, Heimbach, unpubl.). Females of the coccinellid *Adalia bipunctata* were reluctant to eat their own eggs but males showed no restraint in eating eggs they had sired (Agarwala & Dixon 1993a). Other adults or larvae may also eat eggs (Dixon 1959, Pienkowski 1965, Witter 1969, Shepard et al. 1973, Ruzicka et al. 1981) or cannibalism may occur between individuals of the same growth stage (Bhattacharyya 1962, Witter 1969, Canard & Duelli 1984, Morris & Cloutier 1987, Sunderland et al. 1994). In the latter case early-moulting larvae may kill other slower-developing larvae while they are still inactive just prior to moulting (Witter 1969, Canard & Duelli 1984). In the Coccinellidae egg cannibalism can be positively correlated with the duration of egg hatch (Ng 1986b). Locating eggs on the end of long egg stalks reduces egg cannibalism by larvae in the Chrysopidae, but, when very hungry, the larvae will climb the stalks and eat the eggs (Canard & Duelli 1984). Infertile eggs within spider eggsacs may be eaten by the hatchlings (Nentwig 1987) and this can provide enough nutrition for them to moult to the third instar before leaving the eggsac (Valerio 1974). A high percentage of coccinellid eggs can be infertile (Pienkowski 1965) providing food for hatchling larvae (Pienkowski 1965, Rogers et al. 1972b, Kawai 1978) which enhances survival and accelerates development (Kawai 1978). Enhanced survival (Canard & Duelli

1984, Ng 1986a,b, Agarwala & Dixon 1992) might also enable the hatchlings to search a larger area for prey before their energy reserves are depleted, but, on the other hand, Pienkowski (1965) found that cannibalistic coccinellid larvae were less active and less dispersive than non-cannibals, thus reducing their impact on pest populations.

Sibling cannibalism can be viewed as an indirect means of converting maternal tissue into offspring tissue; instead of producing a few large eggs the female produces many small eggs and well-nourished larvae result from sibling cannibalism (Canard & Duelli 1984).

Egg and hatchling quality

Egg size, and the proportion that are viable, can be affected by temperature and by the nutritional and mating history of the mother, and by her size and age. Few generalisations are possible because there is enormous variability between species (and sometimes between species within a family) in the extent to which these factors affect egg quality.

Egg viability may decrease with increasing age of the mother (Morris & Cloutier 1987, Nelemans 1987) or with reduced mating frequency (Shepard et al. 1973). Larger eggs are often more viable (Heessen 1980, Kennedy et al. 1986, Heimbach, unpubl.). Viability can also be affected by the mother's diet (Sundby 1967) and mixed diets may improve viability compared with single-species diets (Toft 1995, Toft, this volume). Eggs of the coccinellid *Adalia bipunctata* were of lower viability when the mother ate *Aphis fabae* compared with viability on a diet of other aphid species, but this effect was not noted for *Coccinella 7-punctata* (Blackman 1967, Hämäläinen & Markkula 1972). Older (Edgar 1971b) or larger (Ernsting & Isaaks 1994) females may lay larger eggs, or there may be no relationship between age (Wallin et al. 1992; Sunderland et al. this volume) or size (Juliano 1985, Nelemans 1987) of mother and egg size. Larger eggs may be more frequent in small broods (Luff 1982, Stewart et al. 1991a) or there may be no relationship between egg size and brood size (Fritz & Morse 1985). In the Coccinellidae egg size is not correlated with the number of ovarioles (Stewart et al. 1991b) nor is it affected by food stress (Dixon & Guo 1993) and it is probably constrained by the minimum size at which neonate larvae can capture prey (Dixon & Guo 1993). Larger eggs with a greater yolk content may result when the mother is supplied with more (Turnbull 1962, Spiller 1992), or more varied (Wallin et al. 1992) food and some foods result in a high nitrogen and carbon content of eggs (Van Dijk, this volume), but equally there are many examples to illustrate an absence of such effects (Kessler 1971, Wise 1979, De Keer & Maelfait 1988, O'Neil & Wiedenmann 1990).

Larger eggs may result if the female is kept at temperatures towards the lower end of the range of temperatures normally experienced (Ernsting & Isaaks 1994), but in other cases smaller eggs result (Sunderland et al. this volume). In some species the development rate of large eggs is slower than that of small eggs (Ernsting & Isaaks 1994), but in other species the reverse is true (Stewart et al. 1991a, Wallin et al. 1992). It is thought that selection for large eggs in the Coccinellidae could be to ensure rapid development so that larvae will hatch before aphid abundance declines (Stewart et al. 1991a).

The fitness of the juvenile stages hatching from eggs is often related to egg quality.

In Carabidae and Staphylinidae, the weight of larvae emerging from large eggs, and their subsequent rate of development and survival probability may be greater than for larvae emerging from smaller eggs (Heessen 1980; Wallin et al. 1992). The size of hatchlings of the linyphiid *Erigone atra* depended on the size, age and diet of their mother (Toft 1995; Toft, this volume) and the proportion of spiderlings of the linyphiid *Oedothorax apicatus* surviving to the adult stage decreased with increasing age of their mother (Dinter & Poehling 1995). Amongst parasitoids offspring fitness often declines with increasing clutch size. Greater clutch size may be a tactic to increase the chances that at least some survive host defences, such as encapsulation of eggs (Godfray 1994).

Advantages and disadvantages of habitat diversification

If measures were taken to increase the recruitment rate of natural enemies, and no other changes were made, this would not necessarily increase natural enemy abundance in the case of species having strong density-dependent population regulation (Schaefer 1978, Brunsting & Heessen 1984) as the increase in population size could be followed rapidly by increased mortality or emigration. This will not, however, apply to all species. Also, if reproduction is increased by manipulating the quantity and quality of the food supply (for example by habitat diversification) this is likely to increase the carrying capacity of the habitat and allow survival of the extra recruits. In addition, other aspects of habitat diversification may independently increase the carrying capacity and permit survival of the enhanced natural enemy population. For example, the availability of suitable web-sites for spiders is often a limiting factor and the carrying capacity of a habitat may be increased by promoting a greater structural diversity of vegetation and more varied soil surface structure (Schaefer 1978, Hatley & MacMahon 1980, Rypstra 1983, Greenstone 1984, Riechert 1990, Alderweireldt 1994, White & Hassall 1994, Heimbach & Garbe, this volume).

Habitat diversification may not be the best strategy for promoting biological control of some pest species. Pests, such as the cabbage root fly (*Delia radicum*), that are available to natural enemies for a very limited period of time before finding protection within plant structures, may be better controlled by specialist natural enemies adapted to their specific biology (Finch, this volume) rather than by generalist natural enemies. In such cases, active measures should be taken to increase the numbers and effectiveness of a limited number of carefully-chosen specialist natural enemies, and these measures need not necessarily include habitat diversification. However, many pests (such as the majority of Homoptera and Lepidoptera) live and feed on the exterior of plants and are available to natural enemies for most, if not all, of their life. There are arguments for supposing that these exposed pests may be controlled effectively by a complex of, mainly generalist, natural enemies. The abundance of any given single species of natural enemy will fluctuate from year to year (due to the weather, and its own natural enemies etc.) so pest control by a single species of natural enemy is inherently unreliable. Complexes of natural enemies, on the other hand, are buffered from this problem. There are sufficient replacement species within each guild that a deficiency in the numbers of any one species in a given year is very likely to be balanced by a superabundance of another species within the same guild (and especially so if they are in competition). In addition, a natural enemy complex encompasses a wide variety of niches (nocturnal plant searchers, ground-based sit-and-wait predators etc.) and so any change in the biology or distribution of the pest, perhaps resulting from a change in farming practice, is unlikely to enable the pest to escape from attack by at least some of the natural enemy species in the complex. The size of natural enemy complexes in agroecosystems should not be underestimated; there are, for example, more than 600 species of spider in USA field crops (Young & Edwards 1990), *c.* 400 species of generalist predator in UK cereals (Potts & Vickerman 1974, Sunderland et al. 1985) and 600-1000 species of predator in USA cotton and soybean (Gross 1987). This represents a rich biological resource potentially available for pest control, but knowledge is currently lacking about how best to manage this resource to our advantage.

It has become clear from this review that there is no single set of environmental conditions that will favour all species of natural enemy. Under each heading of the review it has become apparent that there is variation, even between closely-related species, in optimal physical and biotic conditions. Different stages in the life cycle of a single species have also been found to vary greatly in their requirements. Therefore to maximise the abundance, diversity and quality of individuals in a natural enemy complex by increasing their reproduction (and also by reducing mortality and retaining individuals within fields) it will be necessary create conditions that offer a wide range of *choice* of food and physical conditions, in other words to promote habitat diversification. Within-field diversification is needed for the less mobile species. However, the majority of agricultural species are r-strategists and are therefore highly mobile. These species will benefit from diversification on a greater spatial scale which can include uncropped habitats adjacent to fields. The aim of habitat diversification is not only to increase regional populations of natural enemies but also to ensure that they are abundantly present in crops early in the growing season to prevent the increase of pests.

Many techniques for implementing habitat diversification have been devised. Measures to promote within-field diversification include polycultures, undersowing, managed weediness, intercropping, cover crops, cultivar mixtures, strip-cutting, mulching, weed strips and beetle banks. Measures that can be taken at the edges of fields include conservation headlands, uncropped wildlife strips, wildflower mixtures, and strips of pollen- and nectar-rich plants (e.g. *Phacelia* or *Sinapis*). More details about these techniques are to be found in recent reviews (e.g. Altieri & Letourneau 1982, Risch et al. 1983, Speight 1983, Sheehan 1986, Russell 1989, Altieri 1991, Van Emden & Wratten 1991, Wratten & Van Emden 1995) and information about their effects on pests and natural enemies is becoming available from long-term research programmes on integrated farming systems (Holland et al. 1994b). In addition, natural enemies could be promoted at the landscape level by arranging favourable crops and other habitats into a spatial mosaic with regard to the migration capacities of the most important groups of natural enemies in any particular region (Burel & Baudry 1995). Although habitat diversification might be expected to increase natural enemy abundance this does not necessarily imply a greater impact on pest populations (Finch, this volume). For example, increased biodiversity in fields will provide polyphagous predators with a greater choice of food and may reduce their impact on some less-preferred pest species (Bilde & Toft 1994, Toft 1995). However, although exceptions can be found, the predominant conclusion from the reviews listed above is that habitat diversification both increases natural enemy abundance and reduces pest populations. The authors of an excellent recent review concluded that simple cost-effective habitat diversification measures are now available that will increase natural abundance and reduce pests (Wratten & Van Emden 1995). Habitat diversification can reduce pest attack directly by reducing the level of stimuli attractive to pests, such as olfactory stimuli from crop volatiles or visual stimuli from bare ground contrasting with crop vegetation (resource concentration hypothesis) or it can reduce pests indirectly by encouraging natural enemies (natural enemies hypothesis) (Root 1973). Few studies have separated unequivocally the contributions from these two mechanisms (Sheehan 1986), but it is now generally recognised that in most crops both mechanisms may operate simultaneously to reduce pest attack (Risch et al. 1983, Russell 1989). Some of the diversification techniques listed above, on the one hand reduce pest attack, but on the other hand also impair the yield or quality of the crop (for example by competing with the crop at an early crop growth stage or by encouraging crop diseases). The challenge now is to find a greater range of diversification techniques that will reduce pest incidence without deleterious side-effects on the crop.

This review has identified some major categories of habitat diversification that are of significance for the reproduction (in the broadest sense) of many natural enemies: 1) abundance and variety of food, to enable natural enemies to find easily an adequate supply of high quality food, 2) a range of soil conditions (e.g. moisture level, degree of compaction) to provide suitable oviposition sites and good conditions for larval and pupal survival for a wide range of species, 3) variation in vegetation and soil surface architecture to maximise spider web-site choice, and 4) variation in degree of insolation at the soil surface so that some species can benefit from a heat gain to accelerate egg development whilst nocturnal species can find a cool moist shelter site during the daytime. Many of these requirements would probably be satisfied by a combination of diversification of vegetation and incomplete mulching within crops plus provision of extra floral resources at the edges of crops. It is clear from this review that little is known about the larval ecology, food preferences and oviposition behaviour of the majority of natural enemies in our crops. Intensified study of these aspects of natural enemy biology would provide the necessary sound foundation on which to build more precise and effective techniques of habitat diversification to promote pest control without excessive use of pesticides.

Acknowledgements

This publication was made possible through an EU Concerted Action "Enhancement, dispersal and population dynamics of beneficial predators and parasitoids in integrated agroecosystems". We are grateful for the help of the HRI Library staff. KDS was funded by the UK Ministry of Agriculture, Fisheries and Food.

References

Agarwala, B.K. & Dixon, A.F.G. 1992. Laboratory study of cannibalism and interspecific predation in ladybirds. *Ecol. Ent.* 17: 303-9.

Agarwala, B.K. & Dixon, A.F.G. 1993a. Kin recognition – egg and larval cannibalism in *Adalia bipunctata* (Coleoptera, Coccinellidae). *Eur. J. Ent.* 90: 45-50.

Agarwala, B.K. & Dixon, A.F.G. 1993b. Why do ladybirds lay eggs in clusters ? *Funct. Ecol.* 7: 541-48.

Alderweireldt, M. 1994. Habitat manipulations increasing spider densities in agroecosystems – possibilities for biological control. *J. Appl. Ent.* 118: 10-16.

Alderweireldt, M. & DeKeer, R. 1988. Comparison of the life cycle history of three *Oedothorax* species (Araneae, Linyphiidae) in relation to laboratory observations. *Proc. XI Eur. Arachnol. Colloqu.*, Berlin 1988, Technische Universität Berlin, 169-77.

Alderweireldt, M. & Lissens, A. 1988. Laboratoriumwaarnemingen van de ontwikkeling en reproductie bij *Oedothorax apicatus* (Blackwall, 1850) *en Oedothorax retusus* (Westring, 1851). *Nieuwsbr. belg. arachnol. Ver.* 9: 19-26.

Altieri, M.A. 1991. How best can we use biodiversity in agroecosystems ? *Outl. Agric.* 20: 15-23.

Altieri, M. A. & Letourneau, D.K. 1982. Vegetation management and biological control in agroecosystems. *Crop Prot.* 1: 405-30.

Anderson, N.H. 1962. Growth and fecundity of *Anthocoris* spp. reared on various prey (Heteroptera: Anthocoridae). *Entomologia exp. appl.* 5: 40-52.

Ashley, T.R. & Gonzalez, D. 1974. Effect of various food substances on longevity and fecundity of *Trichogramma. Environ. Entom.* 3: 169-71.

Baars, M.A. & Van Dijk, T.S. 1984. Population dynamics of two carabid beetles at a Dutch heathland. II. egg production and survival in relation to density. J. *Anim. Ecol.* 53: 389-400.

Banks, C.J. 1955. An ecological study of Coccinellidae (Col.) associated with *Aphis fabae* Scop. on *Vicia faba. Bull. ent. Res.* 46: 561-87.

Banks, C.J. 1956. The distributions of coccinellid egg batches and larvae in relation to numbers of *Aphis fabae* Scop. on *Vicia faba. Bull. ent. Res.* 47: 47-56.

Bartlett, B.R. 1964. Patterns in the host-feeding habit of adult parasitic Hymenoptera. *Ann. ent. Soc. Am.* 57: 344-50.

Bauer, T. 1982. Prey-capture in a ground-beetle larva. *Anim. Behav.* 30: 203-8.

Betz, J.O. 1992. Studies on winter-active larvae of the ground beetle *Carabus problematicus* (Coleoptera, Carabidae). *Pedobiologia* 36: 159-67.

Bhattacharyya, S.K. 1962. Laboratory studies on the feeding habits and life cycles of soil inhabiting mites. *Pedobiologia* 1: 291-98.

Bilde, T. & Toft, S. 1994. Prey preference and egg production of the carabid beetle *Agonum dorsale. Entomologia exp. appl.* 73: 151-56.

Bilde, T. & Toft, S. 1995. Quantifying food limitation of arthropod predators. *Acta Jut.*, in press, this volume.

Blackman, R.L. 1967. The effects of different aphid foods on the development of *Adalia bipunctata* L. and *Coccinella 7-punctata* L. *Ann. appl. Biol.* 59: 207-19.

Boldyrev, M.I., Wilde, W.H.A. & Smith, B.C. 1969. Predaceous coccinellid oviposition responses to *Juniperus* wood. *Can. Ent.* 101: 1199-1206.

Brandmayr, P. & Brandmayr, T.Z. 1979. The evolution of parental care phenomena in pterostichine ground beetles, with special reference to the genera *Abax* and *Mollops* (Coleoptera, Carabidae). In: Den Boer, P.J., Thiele, H.U., Weber, F., Veenman, H. & Zonen, B.V. (eds.) *On the Evolution of Behaviour in Carabid Beetles*, Wageningen, 35-49.

Braman, S.K. & Yeargan, K.V. 1988. Comparison of development and reproductive rates of *Nabis americoferus, N. roseipennis* and *N. rufusculus* (Hemiptera: Nabidae). *Ann. Entomol. Soc. Am.* 81: 923-30.

Braman, S.K. & Yeargan, K.V. 1989. Reproductive strategy of *Trissolcus euschisti* (Hymenoptera: Scelionidae) under conditions of partially used host resources. *Ann. Ent. Soc. Amer.* 82: 172-76.

Braman, S.K. & Yeargan, K.V. 1991. Reproductive strategies of primary parasitoids of the green cloverworm (Lepidoptera: Noctuidae). *Environ. Entom.* 20: 349-53.

Briggs, J.B. 1965. Biology of some ground beetles injurious to strawberries. *Bull. ent. Res.* 56: 79-93.

Brunsting, A.M.H. & Heessen, H.J.L. 1983. Cannibalism, laboratory artefact or natural phenomenon. *Rep. 4th Symp. Carab.* 135-39.

Brunsting, A.M.H. & Heessen, H.J.L. 1984. Density regulation in the carabid beetle *Pterostichus oblongopunctatus. J. Anim. Ecol.* 53: 751-60.

Brunsting, A.M.H., Siepel, H. & van Schaick Zillesen, P.G. 1986. The role of larvae in the population ecology of Carabidae. In: Den Boer, P.J., Luff, M.L., Mossakowski, D. & Weber, F. (eds.) *Carabid Beetles, their Adaptations and Dynamics*, Gustav Fischer, Stuttgart, 399-411.

Budenberg, W.J. 1990. Honeydew as a contact kairomone for aphid parasitoids. *Entomol. exp. appl.* 55: 139-48.

Burel, F. & Baudry, J. 1995. Farming landscapes and insects. In: Glen, D.M., Greaves, M.P. & Anderson, H.M. (eds.) *Ecology and Integrated Farming Systems: Proceedings of the 13th Long Ashton Symposium*, Wiley & Sons, Chichester, 203-20.

Bush, L., Kring, T.J. & Ruberson, J.R. 1993. Suitability of greenbugs, cotton aphids, and *Heliothis virescens* eggs for development and reproduction of *Orius insidiosus. Entomologia exp. appl.* 67: 217-22.

Canard, M. & Duelli, P. 1984. Predatory behaviour of larvae and cannibalism. In: Canard, M., Semeria, Y. & New, T.R. (eds.) *Biology of Chrysopidae*, Dr W. Junk, The Hague, 92-100.

Chambers, R.J. 1991. Oviposition by aphidophagous hoverflies (Diptera: Syrphidae) in relation to aphid density and distribution in winter wheat. In: Polgar, L, Chambers, R.J., Dixon, A.F.G. & Hodek, I. (eds.) *Behaviour and Impact of Aphidophaga*, SPB Academic Publishing bv, The Hague, The Netherlands, 115-21.

Chandler, A.E.F. 1967. Oviposition responses by aphidophagous Syrphidae (Diptera). *Nature* 213: 736.

Chandler, A.E.F. 1968a. Some factors influencing the occurrence and site of oviposition by aphidophagous Syrphidae (Diptera). *Ann. appl. Biol.* 61: 435-46.

Chandler, A.E.F. 1968b. The relationship between aphid infestations and oviposition by aphidophagous Syrphidae (Diptera). *Ann. appl. Biol.* 61: 425-34.

Chandler, A.E.F. 1968c. Some host-plant factors affecting oviposition by aphidophagous Syrphidae (Diptera). *Ann. appl. Biol.* 61: 415-23.

Chandler, A.E.F. 1969. Locomotory behaviour of first instar larvae of aphidophagous Syrphidae after contact with aphids. *Anim. Behav.* 17: 673-78.

Chiverton, P.A. & Sotherton, N.W. 1991. The effects on beneficial arthropods of the exclusion of herbicides from cereal crop edges. *J. appl. Ecol.* 28: 1027-39.

Declercq, P. & Degheele, D. 1992. Influence of feeding interval on reproduction and longevity of *Podisus sagitta* (Het., Pentatomidae). *Entomophaga* 37: 583-90.

De Keer, R., Alderweirelt, M., Decleer, K., Segers, H., Desender, K. & Maelfait, J.P. 1989. Horizontal distribution of the spider fauna of intensively grazed pastures under the influence of diurnal activity and grass height. *J. appl. Ent.* 107: 455-73.

De Keer, R. & Maelfait, J.P. 1987. Laboratory observations on the development and reproduction of *Oedothorax fuscus* (Blackwall, 1834) (Araneida, Linyphiidae) under different conditions of temperature and food supply. *Rev. Écol. Biol. Sol.* 24: 63-73.

De Keer, R. & Maelfait, J.P. 1988. Laboratory observations on the development and reproduction of *Erigone atra* Blackwall, 1833 (Araneae, Linyphiidae). *Bull. Br. arachnol. Soc.* 7: 237-42.

Dennis, P., Wratten, S.D. & Sotherton, N.W. 1990. Feeding behaviour of the staphylinid beetle *Tachyporus hypnorum* in relation to its potential for reducing aphid numbers in wheat. *Ann. appl. Biol.* 117: 267-76.

Dennis, P., Wratten, S.D. & Sotherton, N.W. 1991. Mycophagy as a factor limiting predation of aphids (Hemiptera: Aphididae) by staphylinid beetles (Coleoptera: Staphylinidae) in cereals. *Bull. Ent. Res.* 81: 25-31.

Desender, K., van den Broeck, D. & Maelfait, J.P. 1985. Population biology and reproduction in *Pterostichus melanarius* Ill. (Coleoptera, Carabidae) from a heavily grazed pasture ecosystem. *Med. Fac. Landbouww. Rijksuniv. Gent* 50: 567-75.

Dinter, A. & Poehling, H.M. 1995. Side-effects of insecticides on two erigonid spider species. *Entomologia exp. appl.* 74: 151-63.

Dixon, A.F.G. 1959. An experimental study of the searching behaviour of the predatory coccinellid beetle *Adalia decempunctata* (L.). *J. Anim. Ecol.* 28: 259-81.

Dixon, A.F.G. & Guo, Y.Q. 1993. Egg and cluster-size in ladybird beetles (Coleoptera, Coccinellidae) – the direct and indirect effects of aphid abundance. *Eur. J. Ent.* 90: 457-63.

Edgar, W.D. 1971a. Aspects of the ecology and energetics of the eggsac parasites of the wolf spider *Pardosa lugubris* (Walck.). *Oecologia* 7: 155-63.

Edgar, W.D. 1971b. Seasonal weight changes, age structure, natality and mortality in the wolf spider *Pardosa lugubris* Walck. in Central Scotland. *Oikos* 22: 84-92.

Edwards, R.L. 1954. The effect of diet on egg maturation and resorbtion in *Mormoniella vitripennis* (Hymenoptera, Pteromalidae). *Quart. J. Micr. Sci.* 95: 459-68.

Eghtedar, E. 1970. Zur Biologie und Ökologie der Staphyliniden *Philonthus fuscipennis* Mannh. und *Oxytelus rugosus* Grav. *Pedobiologia* 10: 169-79.

Eisner, T. & Eisner, M. 1991. Unpalatability of the pyrrolizidine alkaloid-containing moth *Utetheisa ornatrix*, and its larva, to wolf spiders. *Psyche* 98: 111-18.

Ernsting, G. & Huyer, F.A. 1984. A laboratory study on temperature relations of egg production and development in two related species of carabid beetle. *Oecologia* 62: 361-67.

Ernsting, G. & Isaaks, J.A. 1994. Egg size variation in *Notiophilus biguttatus* (Col., Carabidae). In: Desender, K., Dufrêne, M., Loreau, M., Luff, M.L. & Maelfait, J.P. (eds.) *Carabid Beetles: Ecology and Evolution*, Kluwer Academic Publishers, Dordrecht, 133-37.

Ernsting, G., Isaaks, J.A. & Berg, M.P. 1992. Life cycle and food availability indices in *Notiophilus biguttatus* (Coleoptera, Carabidae). *Ecol. Ent.* 17: 33-42.

Evans, E.W. 1982. Consequence of body size for fecundity in the predatory stinkbug, *Podisus maculiventris* (Hemiptera: Pentatomidae). *Ann. Entomol. Soc. Am.* 75: 418-20.

Finch, S. 1995. Enhancement of beneficial predators and parasitoids in field vegetable crops. *Acta Jut.*, in press, this volume.

Flanders, S.E. 1942. Oösorption and ovulation in relation to oviposition in the parasitic Hymenoptera. *Ann. Ent. Soc. Amer.* 35: 251-66.

Foster, M.A. & Ruesink, W.G. 1984. Influence of flowering weeds associated with reduced tillage in corn on a black cutworm (Lepidoptera: Noctuidae) parasitoid, *Meterus rubens* (nees von Esenbeck). *Environ. Entom.* 13: 664-68.

Frank, J.H. 1968. Notes on the biology of *Philonthus decorus* (Grav.)(Col., Staphylinidae). *Entomologist's mon. Mag.* 103: 273-77.

Fritz, R.S. & Morse, D.H. 1985. Reproductive success and foraging in the crab spider *Misumena vatia*. *Oecologia*, 65: 194-200.

Gauld, I. & Bolton, B. (eds.) 1988. *The Hymenoptera.* Oxford University Press, Oxford. 332 pp.

Ghanim, A.E.B., Freier, B. & Wetzel, T. 1984. Zur Nahrungsaufnahme und Eiablage von *Coccinella septempunctata* L. bei unterschiedlichen Angebot von Aphiden der Arten *Macrosiphum avenae* (Fabr.) und *Rhopalosiphum padi* (L.). *Arch. Phytopathol. u. Pflanzenschutz* 20: 117-25.

Godfray, H.C.H. 1994. *Parasitoids.* Intercept Ltd, Andover, 520 pp.

Good, J.A. 1982. Notes on biogeography and ecology of the common earwig, *Forficula auricularia* (Dermaptera) in Ireland. Part 2. Life cycle. *Ir. Nat. J.* 20: 543-46.

Good, J. A. & Giller, P.S. 1988. A contribution to a check-list of Staphylinidae (Coleoptera) of potential importance in the integrated protection of cereal and grass crops. In: Cavalloro, R. & Sunderland, K.D. (eds.) *Integrated Crop Protection in Cereals*, A.A. Balkema, Rotterdam, 81-98.

Greenstone, M.H. 1979. Spider feeding behaviour optimises dietary essential amino acid composition. *Nature* 282: 501-3.

Greenstone, M.H. 1984. Determinants of web spider species diversity: vegetation structural diversity vs. prey availability. *Oecologia* 62: 299-304.

Gross, H.R. 1987. Conservation and enhancement of entomophagous insects – a perspective. *J. Entomol. Sci.* 22: 97-105.

Grüm, L. 1973. Egg production of some Carabidae species. *Bull. Acad. Pol. Sci. Biol.* 21: 261-68.

Hackman, W. 1957. Studies on the ecology of the wolf spider *Trochosa ruricola* (Deg.). *Soc. Sci. Fenn. Commentat. Biol.* 16: 1-34.

Hallander, H. 1970. Prey, cannibalism and microhabitat selection in the wolf spiders *Pardosa chelata* O.F. Muller and *P. pullata* Clerck. *Oikos* 21: 337-40.

Hämäläinen, M. & Markkula, M. 1972. Effect of type of food on fecundity in *Coccinella septempunctata* L. (Col., Coccinellidae). *Ann. Ent. Fenn.* 38: 195-99.

Hämäläinen, M., Markkula, M. & Raij, T. 1975. Fecundity and larval voracity of four ladybeetle species. *Ann. Ent. Fenn.* 41: 124-27.

Hamed, A.R. 1979. Zur Wirkung von *Bacillus thuringiensis* auf Parasiten und Predatoren von *Yponomeuta evonynellus. Z. ang. Ent.* 87: 294-311.

Hatley, C.L. & MacMahon, J.A. 1980. Spider community organisation; seasonal variation and the role of vegetation architecture. *Env. Ent.* 9: 632-39.

Heessen, H.J.L. 1980. Egg production in *Pterostichus oblongopunctatus* (Fab.) (Col., Carabidae) and *Philonthus decorus* (Grav.)(Col., Staphylinidae). *Neth. J. Zool.* 30: 30-53.

Heesen, H.J.L. 1981. Egg mortality in *Pterostichus oblongopunctatus* (F.) (Col., Carabidae). *Oecologia*, 50: 233-35.

Heessen, H.J.L. & Brunsting, A.M.H. 1981. Mortality of larvae of *Pterostichus oblongopunctatus* (Fabricius)(Col., Carabidae) and *Philonthus decorus* (Gravenhorst)(Col., Staphylinidae). *Neth. J. Zool.* 31: 729-45.

Heimbach, U. 1989. Massenzucht von *Poecilus cupreus* (Col., Carabidae). *J. Gesellsch. Ökol. Osnabrück* 19: 228-29.

Heimbach, U. & Garbe, V. Effects of reduced tillage systems in sugar beets on predatory and pest arthropods. *Acta Jut.*, in press, this volume.

Heimpel, G.E. & Rosenheim, J.A. 1995. Dynamic host feeding by the parasitoid *Aphytis melinus*: the balance between current and future reproduction. *J. Anim. Ecol.* 64: 153-67.

Hemptinne, J.L., Naisse, J. & Os, S. 1988. Glimpse of the life history of *Propylea quatuordecimpunctata* (L.)(Coleoptera: Coccinellidae). *Med. Fac. Landbouww. Rijksuniv. Gent* 53: 1175-82.

Hemptinne, J.L., Dixon, A.F.G. & Coffin, J. 1992. Attack strategy of ladybird beetles (Coccinellidae): factors shaping their numerical response. *Oecologia* 90: 238-45.

Hemptinne, J.L., Dixon, A.F.G, Doucet, J.L. & Petersen, J.E. 1993. Optimal foraging by hoverflies (Diptera: Syrphidae) and ladybirds (Coleoptera: Coccinellidae): Mechanisms. *Eur. J. Ent.* 90: 451-55.

Hodek, I. 1957. The influence of *Aphis sambuci* L. as food for *Coccinella 7-punctata* L. II. *Acta Soc. ent. Bohem. (Csl.)* 54: 10-17.

Hodek, I. 1993. Habitat and food specificity in aphidophagous predators. *Biocont. Sci. Tech.* 3: 91-100.

Holland, J.M., Frampton, G.K., Çigli, T. & Wratten, S.D. 1994b. Arable acronyms analysed – a review of integrated arable farming systems research in Western Europe. *Ann. appl. Biol.* 125: 399-438.

Holmberg, R.G. & Turnbull, A.J. 1982. Selective predation in a euryphagous invertebrate predator, *Pardosa vancouveri* (Arachnida: Araneae). *Can. Ent.* 114: 243-57.

Humphreys, W.F. 1980. Comment on M.H. Greenstone's paper "Spider feeding optimises dietary essential amino acid composition" *Nature* 284: 578.

Hůrka, K. 1975. Laboratory studies of the life cycle of *Pterostichus melanarius* (Illig.) (Coleoptera: Carabidae). *Věst. Cesk. Spol. Zool.* 39: 265-74.

Idoine, K. & Ferro, D.N. 1988. Aphid honeydew as a carbohydrate source for *Edovum puttleri* (Hymenoptera: Eulophidae). *Environ. Entom.* 17: 941-44.

Iperti, G. & Prudent, P. 1986. Effect of the substrate properties on the choice of oviposition sites by *Adalia bipunctata*. In: Hodek, I. (ed.) *Ecology of Aphidophaga 2*, Academia, Prague, 143-49.

Iperti, G. & Quilici, S. 1986. Some factors influencing the selection of oviposition site by *Propylea quatuordecimpunctata*. In: Hodek, I. (ed.) *Ecology of Aphidophaga 2*, Academia, Prague, 137-42.

Jepson, P.C. 1989. *Pesticides and Non-target Invertebrates*. Intercept, Dorset, UK, 240 pp.

Jervis, M.A. & Kidd, N.A.C. 1986. Host-feeding strategies in hymenopteran parasitoids. *Biol. Rev.* 61: 395-434.

Jervis, M.A., Kidd, N.A.C. & Walton, M. 1992. A review of methods for determining dietary range in adult parasitoids. *Entomophaga* 37: 565-74.

Jervis, M.A., Kidd, N.A.C., Fitton, M.G., Huddleston, T. & Dawah, H.A. 1993. Flower-visiting by hymenopteran parasitoids. *J. Nat. Hist.* 27: 67-105.

Juliano, S.A. 1985. The effects of body size on mating and reproduction in *Brachinus lateralis* (Coleoptera: Carabidae). *Ecol. Ent.* 10: 271-80.

Juliano, S.A. 1986. Food limitation of reproduction and survival for populations of *Brachinus* (Coleoptera, Carabidae). *Ecology* 67: 1036-45.

Kan, E. 1989. Assessment of aphid colonies by hoverflies. III. Pea aphids and *Episyrphus balteatus* (De Geer) (Diptera: Syrphidae). *J. Ethol.* 7: 1-6.

Kawai, A. 1978. Sibling cannibalism in the first instar larvae of *Harmonia axyridis* Pallas (Coleoptera, Coccinellidae). *Kontyu* 46: 14-19.

Kenmore, P.E., Carlno, F.O., Perez, C.A., Dyck, V.A. & Gutierrez, A.P. 1984. Population regulation of the Rice Brown Planthopper (*Nilaparvata lugens* Stal.) within rice fields in the Philippines. *J. Pl. Prot. Tropics* 1: 19-37.

Kennedy, T.F., Evans, G.O. & Feeney, A.M. 1986. Studies on the biology of *Tachyporus hypnorum* F. (Col. Staphylinidae) associated with cereal fields in Ireland. *Ir. J. agric. Res.* 25: 81-95.

Kessler, A.M. 1971. Relation between egg production and food consumption in species of the genus *Pardosa* (Lycosidae, Araneae) under conditions of food abundance and food shortage. *Oecologia* 8: 93-109.

Kessler, A. 1973. A comparative study of the production of eggs in eight *Pardosa* species in the field (Araneae, lycosidae). *Tijds. Entomol.* 116: 23-41.

Kessler, A. & Fokkinga, A. 1973. Hymenopterous parasites in egg sacs of spiders of the genus Pardosa (Araneida, Lycosidae). *Tijds. Entomol.* 116: 43-61.

Kiman, Z.B. & Yeargan, K.V. 1985. Development and reproduction of the predator *Orius insidiosus* (Hemiptera: Anthocoridae) reared on diets of selected plant material and arthropod prey. *Ann. Ent. Soc. Amer.* 78: 464-67.

King, P.E. 1963. The rate of egg resorption in *Nasonia vitripennis* (Walker) (Hymenoptera: Pteromalidae) deprived of hosts. *Proc. R. Entomol. Soc. Lond.* (A) 38: 98-100.

King, P.E. & Richards, J.G. 1968. Oösorption in *Nasonia vitripennis* (Hymenoptera: Pteromalidae). *J. Zool.* 154: 495-516.

Krause, U. & Poehling, H.M. 1995. Population dynamics of hoverflies (Diptera: Syrphidae) in Northern Germany in relation to different habitat structure. *Acta Jut.*, in press, this volume.

Leius, K. 1960. Attractiveness of different foods and flowers to the adults of some hymenopterous parasites. *Ca. Ent.* 92: 369-76.

Leius, K. 1961a. Influence of food on fecundity and longevity of adults of *Itoplectis conquisitor* (Say)(Hymenoptera: Ichneumonidae). *Can. Ent.* 93: 771-80.

Leius, K. 1961b. Influence of various foods on fecundity and longevity of adults of *Scambus buolianae* (Htg.)(Hymenoptera: Ichneumonidae). *Can. Ent.* 93: 1079-84.

Leius, K. 1962. Effects of the body fluids of various host larvae on fecundity of females of *Scambus buolianae* (Htg.)(Hymenoptera: Ichneumonidae). *Can. Ent.* 94: 1078-82.

Leius, K. 1963. Effects of pollens on fecundity and longevity of adult *Scambus buolianae* (Htg.)(Hymenoptera: Ichneumonidae). *Can. Ent.* 95: 202-7.

Leius, K. 1967. Food sources and preferences of adults of a parasite, *Scambus buolianae* (Hym.: Ichn.) and their consequences. *Can. Ent.* 99: 865-87.

Lipkow, E. 1966. Biologisch-ökologisch Untersuchungen uber *Tachyporus*-Arten und *Tachinus rufipes* (Col., Staphyl.). *Pedobiologia* 6: 140-77.

Lipkow, E. 1968. Zum Eiablage-Verhalten der Staphyliniden. *Pedobiologia* 8: 208-13.

Luff, M.L. 1982. Population dynamics of Carabidae. *Ann. appl. Biol.* 101: 164-70.

Luff, M.L. 1987. Biology of polyphagous ground beetles in agriculture. *Agric. Zool. Rev.* 2: 237-78.

Luff, M.L. 1994. Starvation capacities of some carabid larvae. In: Desender, K. et al., *Carabid Beetles: Ecology and Evolution*, Kluwer Academic Publishers, The Netherlands, 171-75.

McMurtry, J.A. & Scriven, G.T. 1964. Studies on the feeding, reproduction and development of *Amblyseius hibisci* (Acarina: Phytoseiidae) on various food substances. *Ann. Entomol. Soc. Am.* 57: 649-55.

Metge, K. & Heimbach, U. 1994. Entwicklung eines Zuchtverfahrens für den Staphyliniden *Philonthus cognatus* Steph. *Mitt. a. d. Biol. Bundesanst.* 301: 512.

Mills, N.J. 1981. Essential and alternative foods for some British Coccinellidae (Coleoptera). *Ent. Gaz.* 32: 197-202.

Mills, N.J. 1982. Voracity, cannibalism and coccinellid predation. *Ann. appl. Biol.* 101: 144-48.

Miyashita, K. 1968. Growth and development of *Lycosa T-insignita* Boes. et Str. (Araneae: Lycosidae) under different feeding conditions. *Appl. Ent. Zool.* 3: 81-88.

Miyashita, T. 1992. Food limitation of population density in the orb-web spider, *Nephila clavata*. *Res. Popul. Ecol.* 34: 143-53.

Mols, P.J.M. 1988. Simulation of hunger, feeding and egg production in the carabid beetle *Pterostichus coerulescens* L. (= *Poecilus versicolor* Sturm). *Agric. Univ. Wageningen Papers* 88-3: 99 pp.

Morris, D.E. & Cloutier, C. 1987. Biology of the predatory fly *Coenosia tigrina* (Fab.) (Diptera, Anthomyiidae) – reproduction, development, and larval feeding on earthworms in the laboratory. *Can. Ent.* 119: 381-93.

Murdoch, W.W. 1969. Switching in general predators: experiments on predator specificity and stability of prey populations. *Ecol. Monogr.* 39: 335-54.

Nelemans, M.N.E. 1987. On the life history of the carabid beetle *Nebria brevicollis* (F.). *Neth. J. Zool.* 37: 26-42.

Nelemans, M.N.E. 1988. Surface activity and growth of larvae of *Nebria brevicollis* (F.)(Coleoptera, Carabidae). *Neth. J. Zool.* 38: 74-95.

Nelemans, M.N.E., den Boer, P.J. & Spee, A. 1989. Recruitment and summer diuapause in the dynamics of a population of *Nebria brevicollis* (Coleoptera: Carabidae). *Oikos* 56: 157-69.

Nentwig, W. 1985. Spiders eat crickets artificially poisoned with KCN and change the composition of their digestive fluid. *Naturw.* 72: 545-46.

Nentwig, W. 1987. The prey of spiders. In: Nentwig, W. (ed.) *Ecophysiology of Spiders*, Springer-Verlag, Berlin, 249-63.

Nettles, W.C. 1987. *Eucelatoria bryani* (Diptera: Tachinidae): Effect on fecundity of feeding on hosts. *Environ. Entom.* 16: 437-40.

Ng, S.M. 1986a. Effects of sibling egg cannibalism on the first instar larvae of four species of aphidophagous Coccinellidae. In: Hodek, I. (ed.) *Ecology of Aphidophaga 2*, Academia, Prague, 69-75.

Ng, S.M. 1986b. Egg mortality of four species of aphidophagous Coccinellidae in Malaysia. In: Hodek, I. (ed.) *Ecology of Aphidophaga 2*, Academia, Prague, 77-81.

O'Neil, R.J. 1992. Body weight and reproductive status of 2 nabid species (Heteroptera, Nabidae) in Indiana. *Environ. Entom.* 21: 191-96.

O'Neil, R.J. & Wiedenmann, R.N. 1990. Body weight of *Podisus maculiventris* (Say) under various feeding regimens. *Can. Ent.* 122: 285-94.

Orr, D.B., Boethel, D.J. & Jones, W.A. 1985. Biology of *Telenomus chloropus* (Hymenoptera: Scelionidae) from eggs of *Nezara viridula* (Hemiptera: Pentatomidae) reared on resistant and susceptible soybean genotypes. *Can. Ent.* 117: 1137-42.

Ouyang, Y., Grafton Cardwell, E.E. & Bugg, R.L. 1992. Effects of various pollens on development, survivorship, and reproduction of *Euseius tularensis* (Acari, Phytoseiidae). *Environ. Entom.* 21: 1371-76.

Palanichamy, S. 1985. Effect of temperature on food utilisation, growth and egg production in the spider *Cyrtophora cicatrosa*. *J. Therm. Biol.* 10: 63-70.

Pienkowski, R.L. 1965. The incidence and effect of egg cannibalism in first-instar *Coleomegilla maculata lengi* (Coleoptera: Coccinellidae). *Ann. Entomol. Soc. Am.* 58: 150-53.

Pimentel, D. 1995. Ecological theory, pest problems and biologically based solutions. In: Glen, D.M., Greaves, M.P. & Anderson, H.M. (eds.) *Ecology and Integrated Farming Systems: Proceedings of the 13th Long Ashton Symposium*, Wiley & Sons, Chichester, 69-82.

Potts, G.R. & Vickerman, G.P. 1974. Studies on the cereal ecosystem. *Adv. Ecol. Res.* 8: 107-97.

Powell, W. 1986. Enhancing parasitoid activity in crops. In: Waage, J. & Greathead, D. (eds.) Insect Parasitoidseds. *13th Symp. R. Entomol. Soc. Lond.* Academic Press, London, 319-40.

Riechert, S.E. 1990. Habitat manipulations augment spider control of insect pests. *Acta Zool. Fennica* 190: 321-25.

Risch, S.J., Andow, D. & Altieri, M.A. 1983. Agroecosystem diversity and pest control: data, tentative conclusions and new research directions. *Env. Entom.* 12: 625-29.

Rollard, C. 1990. Mortality of spider eggs in Brittany. *Acta Zool. Fennica* 190: 327-31.

Rogers, C.E., Jackson, H.B & Eikenbary, R.D. 1972a. Responses of an imported coccinellid, *Propylea 14-punctata* to aphids associated with small grains in Oklahoma. *Environ. Entom.* 1: 198-202.

Rogers, C.E., Jackson, H.B, Angalet, G.W. & Eikenbary, R.D. 1972b. Biology and life history of *Propylea 14-punctata* (Coleoptera: Coccinellidae) an exotic predator of aphids. *Ann. ent. Soc. Am.* 65: 648-50.

Root, R.B. 1973. Organisation of a plant-arthropod association in simple and diverse habitats: the fauna of collards (*Brassica oleracea*). *Ecol. Monogr.* 43: 95-124.

Ruberson, J.R., Young, S.Y. & Kring, T.J. 1991. Suitability of prey infected by nuclear polyhedrosis virus for development, survival, and reproduction of the predator *Nabis roseipennis* (Heteroptera: Nabidae). *Environ. Entomol.* 20: 1475-79.

Russell, E.P. 1989. Enemies hypothesis: a review of the effects of vegetational diversity on predatory pests and parasitoids. *Env. Entomol.* 18: 590-99.

Ruth, J. & Dwumfour, E.F. 1989. Laboruntersuchungen zur Eignung einiger Blattlausarten als Beute der räuberischen Blumenwanze *Anthocoris gallarum-ulmi* (DeG.) (Het., Anthocoridae). *J. appl. Ent.* 108: 321-27.

Ruzicka, Z., Iperti, G. & Hodek, I. 1981. Reproductive rate and longevity in *Semiadalia undecimnotata* and *Coccinella septempunctata* (Coccinellidae, Col.). *Vest. cs. Spolec. zool.* 45: 115-28.

Rypstra, A.L. 1983. The importance of food and space in limiting web-spider densities; a test using field enclosures. *Oecologia* 59: 312-16.

Sahragard, A., Jervis, M.A. & Kidd, N.A.C. 1991. Influence of host availability on rates of oviposition and host feeding and on longevity in *Dicondylus indianus* Olmi (Hym., Dryinidae), a parasitoid of the Rice Brown Planthopper, *Nilaparvata lugens* Stal. (Hem., Delphacidae). *J. appl. Ent.* 112: 153-62.

Salama, H.S., Zaki, F.N. & El-Sharaby, A.F. 1982. Effect of *Bacillus thuringiensis* Berliner on parasites and predators of the cotton leafworm *Spodoptera littoralis* (Boisd.). *Z. ang. Ent.* 94: 498-504.

Salama, H.S., El-Moursy, A., Zaki, F.N., Abou-Ela, R. & Abdel-Razek, A. 1991. Parasites and predators of the meal moth *Plodia interpunctella* Hbn. as affected by *Bacillus thuringiensis* Berl. *J. appl. Ent.* 112: 244-53.

Sandlan, K.P., Jones, R.L. & Chiang, H.C. 1983. Influence of density of the European corn borer, Ostrinia nubilalis (Lepidoptera: Pyralidae) on the parasitoid Lydella thompsoni (Diptera: Tachinidae). *Environ. Entom.* 12: 174-77.

Schaefer, M. 1977. Untersuchungen über das Wachstum von zwei Spinnenarten (Araneida) im Labor und Freiland. *Pedobiologia* 17: 189-200.
Schaefer, M. 1978. Some experiments on the regulation of population density in the spider *Floronia bucculenta* (Araneida:Linyphiidae) *Symp. Zool. Soc.* 42: 203-10.
Schelvis, J. & Siepel, H. 1988. Larval food spectra of *Pterostichus oblongopunctatus* and *P. rhaeticus* in the field. *Entomol. Gener.* 13: 61-66.
Sengonca, C., Gerlach, S. & Melzer, G. 1987. Einfluss der Ernährung mit unterschiedlicher Beute auf *Chrysoperla carnea* (Stephens) (Neuroptera: Chrysopidae). *Z. Pflanzenkr. Pflanzensch.* 94: 197-205.
Sheehan, W. 1986. Response by specialist and generalist natural enemies to agroecosystem diversification: a selective review. *Env. Entomol.* 15: 456-61.
Shepard, M., Waddill, V. & Kloft, W. 1973. Biology of the predaceous earwig, *Labidura riparia* (Dermaptera: Labiduridae). *Ann. ent. Soc. Amer.* 66: 837-41.
Speight, M.R. 1983. The potential of ecosystem management for pest control. *Agric. Ecos. & Env.* 10: 183-99.
Sota, T. 1985. Activity patterns and interspecific interactions of coexisting spring and autumn breeding carabids: *Carabus yaconicus* and *Leptocarabus kumagaii* (Coleoptera, Carabidae). *Ecol. Ent.* 10: 315-24.
Sota, T. 1985b. Limitation of reproduction by feeding condition in a carabid beetle, *Carabus yaconicus. Res. Popul. Ecol.* 27: 171-84.
Spence, J.R. & Sutcliffe, J.F. 1982. Structure and function of feeding in larvae of *Nebria. Can. J. Zool.* 60: 2382-94.
Spiller, D.A. 1992. Numerical response to prey abundance by *Zygiella-x-notata* (Araneae, Araneidae). *J. Arachnol.* 20: 179-88.
Stary, P. 1970. *Biology of aphid parasites (Hymenoptera: Aphidiidae) with respect to integrated control.* Dr W. Junk, the Hague, 643 pp.
Steigen, A.L. 1975. Energetics in a population of *Pardosa palustris* (L.)(Araneae, Lycosidae) on Hardangervidda. In: Wielgokski, F. (ed.) *Fennoscandian Tundra Ecosystems. Part 2. Animals and Systems Analysis*, Springer, Berlin, 129-44.
Stewart, L.A., Hemptinne, J.L. & Dixon, A.F.G. 1991a. Reproductive tactics of ladybird beetles: relationships between egg size, ovariole number and developmental time. *Funct. Ecol.* 5: 380-85.
Stewart, L.A., Dixon, A.F.G., Ruzicka, Z. & Iperti, G. 1991b. Clutch and egg size in ladybird beetles. *Entomophaga* 36: 93-97.
Sundby, R.A. 1967. Influence of food on the fecundity of *Chrysopa carnea* Stephens (Neuroptera, Chrysopidae). *Entomophaga* 12: 475-79.
Sunderland, K.D. 1992. Effects of pesticides on the population ecology of polyphagous predators. *Asp. Appl. Biol.* 31: 19-28.
Sunderland, K.D., Chambers, R.J., Stacey, D.L. & Crook, N.E. 1985. Invertebrate polyphagous predators and cereal aphids. *Bull. SROP/WPRS VIII/3*, 105-14.
Sunderland, K.D., Crook, N.E., Stacey, D.L. & Fuller, B.J. 1987. A study of feeding by polyphagous predators on cereal aphids using ELISA and gut dissection. *J. appl. Ecol.* 24: 907-33.
Sunderland, K.D., Ellis, S.J., Weiss, A., Topping, C.J. & Long, S.J. 1994. The effects of polyphagous predators on spiders and mites in cereal fields. *Brighton Crop Prot. Conf. 1994*, 1151-56.

Sunderland, K.D., Topping, C.J., Ellis, S., Long, S., van de Laak, S. & Else, M. 1996. Reproduction and survival of linyphiid spiders, with special reference to *Lepthyphantes tenuis* (Blackwall). *Acta Jut.*, in press, this volume.

Suzuki, Y. & Kiritani, K. 1974. Reproduction of *Lycosa pseudoannulata* (Boesenberg and Strand) (Araneae: Lycosidae) under different feeding conditions. *Jap. J. Appl. Ent. Zool.* 18: 166-70.

Syme, P.D. 1975. The effects of flowers on the longevity and fecundity of two native parasites of the European pine shoot moth in Ontario. *Environ, Entom.* 4: 337-46.

Syme, P.D. 1977. Observations on the longevity and fecundity of *Orgilus obscurator* (Hymenoptera: Braconidae) and the effects of certain foods on longevity. *Can. Ent.* 109: 995-1000.

Szyszko, J. Survival and reproduction in relation to habitat quality and food availability for *Pterostichus oblongopunctatus* F. (Carabidae, Col.). *Act. Jut.*, this volume.

Temerak, S.A. 1980. Detrimental effects of rearing a braconid parasitoid on the pink borer larvae inoculated by different concentrations of the bacterium *Bacillus thuringiensis* Berliner. *Z. ang. Ent.* 89: 315-19.

Thang, M.H., Mochida, O. & Morallo-Rejesus, B. 1990. Mass production of the wolf spider, *Lycosa pseudoannulata* (Araneae, Lycosidae) a predator of insect pests, especially hoppers on rice. *Food & Fert. Tech. Center Asian Pacific Region* 40: 199-206.

Theiling, K.M. & Croft, B.A. 1989. Toxicity, selectivity and sublethal effects of pesticides on arthropod natural enemies: a database summary. In: Jepson, P.C. (ed.) *Pesticides and Non-target Invertebrates*, Intercept, Wimbourne, Dorset, pp 213-32.

Theiss, S. & Heimbach, U. 1993. Fütterungsversuche an Carabidenlarven als Beitrag zur Klärung ihrer Biologie. *Mitt. Dtsch. Ges. Allg. Angew. Ent.* 8: 841-47.

Theiss, S. & Heimbach, U. 1994a. Präimaginale Larvalentwicklung der Laufkäfer-Art *Poecilus cupreus* in Abhängigkeit von Bodenfeuchte und Temperatur (Coleoptera: Carabidae). *Entomol. Gener.* 19: 57-60.

Theiss, S. & Heimbach, U. 1994b. Verwendung chemisch konservierter Nahrung zur Laborzucht der Laufkäfer-Art *Poecilus cupreus* (Coleoptera: Carabidae). *Entomol. Gener.* 18: 273-78.

Thiele, H.U. 1977. Carabid Beetles in their Environments. Springer-Verlag, Berlin.

Toft, S. 1979. Life histories of eight Danish wetland spiders. *Ent. Meddr.* 47: 22-32.

Toft, S. 1995. Value of the aphid *Rhopalosiphum padi* as food for cereal spiders. *J. appl. Ecol.* in press.

Toft, S. 1995. Indicators of prey quality for arthropod predators. *Acta Jut.*, in press, this volume.

Topham, M. & Beardsley, J.W. 1975. An influence of nectar source plants on the New Guinea sugarcane weevil parasite, *Lixophaga sphenophori* (Villeneuve). *Proc. Hawaiian Entomol. Soc.* 22: 145-55.

Topping, C.J. & Sunderland, K.D. 1995. Estimation of the mortality rate of eggs and early instar *Lepthyphantes tenuis* (Araneae: Linyphiidae) from measurements of reproduction and development. *Acta Jut.*, in press, this volume.

Treacy, M.F., Benedict, J.H., Walmsley, M.H., Lopez, J.D., & Morrison, R.K. 1987.

Parasitism of bollworm (Lepidoptera: Noctuidae) eggs on nectaried and nectariless cotton. *Environ. Entom.* 16: 420-23.

Turnbull, A.L. 1962. Quantitative studies of the food of *Linyphia triangularis* Clerck (Araneae: Linyphiidae). *Can. Ent.* 94: 1233-49.

Uetz, G.W., Bischoff, J., Raver, J. 1992. Survivorship of wolf spiders (Lycosidae) reared on different diets. *J. Arachnol.* 20: 207-11.

Valerio, C.E. 1974. Feeding on eggs by spiderlings of *Achaearanea tepidariorum* (Araneae: Theridiidae) and the significance of the quiescent instar in spiders. *J. Arachnol.* 2: 57-63.

Van Baarlen, P., Sunderland, K.D. & Topping, C.J. 1994. Parasitism of money spiders (Araneae, Linyphiidae) in cereals, with a simple method for estimating percentage parasitism of *Erigone* spp. eggsacs by Hymenoptera. *J. app. Ent.* 118: 217-23.

Van Dijk, T.S. 1973. The age composition of *Calathus melanocephalus* L. analysed by studying marked individuals kept within fenced sites. *Oecologia* 12: 213-40.

Van Dijk, T.S. 1979. On the relationship between reproduction, age and survival in two carabid beetles: *Calathus melanocephalus* L. and *Pterostichus coerulescens* L. (Coleoptera: Carabidae). *Oecologia* 40: 63-80.

Van Dijk, T.S. 1982. Individual variability and its significance for the survival of animal populations. In: Mossakowski, D. & Roth, G. (eds.) *Environmental Adaptation and Evolution*, Gustav Fischer, Stuttgart, 233-51.

Van Dijk, T.S. 1986. On the relationship between availability of food and fecundity in carabid beetles: how far is the number of eggs in the ovaries a measure of the quantities of food in the field? In: Den Boer, P.J., Grüm, L. & Szyszko, J. (eds.) *Feeding Behaviour and Accessibility of Food for Carabid Beetles*, Warsaw Agricultural University Press, Warsaw, 105-21.

Van Dijk, T.S. 1994. On the relationship between food, reproduction and survival of two carabid beetles: *Calathus melanocephalus* and *Pterostichus versicolor*. *Ecol. Ent.* 19: 263-70.

Van Dijk, T.S. 1995. The influence of environmental factors and food on life cycle, ageing and survival of carabid beetles. *Acta Jut.*, this volume.

Van Dijk, T.S. & Den Boer, P.J. 1992. The life histories and population dynamics of two carabid species on a Dutch heathland. *Oecologia* 90: 340-52.

Van Emden, H.F. 1962. Observations on the effect of flowers on the activity of parasitic Hymenoptera. *Entomologist's mon. Mag.* 98: 265-70.

Van Emden, H.F. & Wratten, S.D. 1991. Tri-trophic interactions involving plants in the biological control of aphids. In: *Aphid-Plant Interactions: Populations to Molecules*, USDA: ARS, Oklahoma State University, 29-43.

Van Lenteren, J.C., van Vianen, A., Gast, H.F. & Kortenhoff, A. 1987. The parasite-host relationship between *Encarsia formosa* Gahan (Hymenoptera: Aphelinidae) and *Trialeurodes vaporariorum* (Westwood) (Homoptera: Aleyrodidae). XVI. Food effects on oogenesis, oviposition, life-span and fecundity of *Encarsia formosa* and other hymenopterous parasites. *J. Appl. Ent.* 103: 69-84.

Van Wingerden, W.K.R.E. 1973. Dynamik einer Population von *Erigone arctica* White (Araneae, Micryphantidae). Prozesse der Natalität. *Faun.-ökol. Mitt.* 4: 207-22.

Van Wingerden, W.K.R.E. 1977. Population dynamics of *Erigone arctica* (White)

(Araneae, Linyphiidae). Thesis, Free University of Amsterdam, The Netherlands.

Van Wingerden, W.K.R.E. 1978. Population dynamics of *Erigone arctica* (White) (Araneae: Linyphiidae). II. *Symp. Zool. Soc. Lond.* 42: 195-202.

Vlijm, L. & Richter, C.J.J. 1966. Activity fluctuations of *Pardosa lugubris* (Walckenaer), Araneae, Lycosidae, during the breeding season. *Entomol. Berichten* 26: 222-30.

Waldbauer, G.P. & Friedman, S. 1991. Self-selection of optimal diets by insects. *A. Rev. Ent.* 36: 43-63.

Wallin, H. 1989. Habitat selection, reproduction and survival of two small carabid species on arable land: a comparison between *Trechus secalis* and *Bembidion lampros*. *Hol. Ecol.* 12: 193-200.

Wallin, H., Chiverton, P.A., Ekbom, B.S. & Borg, A. 1992. Diet, fecundity and egg size in some polyphagous predatory carabid beetles. *Entomologia exp. appl.* 65: 129-40.

Walter, D.E., Hunt, H.W. & Elliott, E.T. 1987. The influence of prey type on the development and reproduction of some predatory soil mites. *Pedobiologia* 30: 419-24.

Weseloh, R.M. (1993). Adult feeding affects fecundity of the predator, *Calosoma sycophanta* (Col.: Carabidae). *Entomophaga* 38: 435-39.

White, T.C.R. 1978. The importance of a relative shortage of food in animal ecology. *Oecologia* 33: 71-86.

White, P.C.L. & Hassall, M. 1994. Effects of management on spider communities of headlands in cereal fields. *Pedobiologia* 38: 169-84.

Willey, M.B. & Adler, P.H. 1989. Biology of *Peucetia viridans* (Araneae, Oxyopidae) in South Carolina, with special reference to predation and maternal care. *J. Arachnol.* 17: 275-84.

Wise, D.H. 1975. Food limitation of the spider *Linyphia marginata*: experimental field studies. *Ecology*, 56: 637-46.

Wise, D.H. 1979. Effects of an experimental increase in prey abundance upon the reproductive rates of two orb-weaving spider species (Araneae: Araneidae). *Oecologia* 41: 289-300.

Witter, J.A. 1969. Laboratory studies on the developmental period and feeding behaviour of *Aphidecta obliterata* (L.) (Coleoptera: Coccinellidae) an introduced predator of the balsam woolly aphid. *Ann. Entomol. Soc. Am.* 62: 1004-08.

Wratten, S.D. 1973. The effectiveness of the coccinellid beetle *Adalia bipunctata* (L.) as a predator of the lime aphid *Eucallipterus tiliae* L. *J. Anim. Ecol.* 42: 785-802.

Wratten, S.D. & Van Emden, H.F. 1995. Habitat management for enhanced activity of natural enemies of insect pests. In: Glen, D.M., Greaves, M.P. & Anderson, H.M. (eds.) *Ecology and Integrated Farming Systems: Proceedings of the 13th Long Ashton Symposium*, Wiley & Sons, Chichester, 117-45.

Wright, E.J. & Laing, J.E. 1978. The effects of temperature on development, adult longevity and fecundity of *Coleomegilla maculata lengi* and its parasite *Perilitus coccinellae*. *Proc. Entomol. Soc. Ont.* 109: 33-47.

Yao, D.S. & Chant, D.A. 1990. Changes in body weight of two species of predatory mites (Acarina: Phytoseiidae) as a consequence of feeding in an interactive system. *Exp. Appl. Acarol.* 8: 195-220.

Young, O.P. & Edwards, G.B. 1990. Spiders in United States field crops and their potential effect on crop pests. *J. Arachnol.* 18: 1-27.

Zangger, A. 1994. The positive influence of strip-management on carabid beetles in a cereal field: accessibility of food and reproduction in *Poecilus cupreus*. In: Desender, K., Dufrêne, M., Loreau, M., Luff, M.L. & Maelfait, J.P. (eds.) *Carabid Beetles: Ecology and Evolution*, Kluwer Academic Publishers, Dordrecht, 469-72.

Zangger, A., Lys, J.A. & Nentwig, W. 1994. Increasing the availability of food and the reproduction of *Poecilus cupreus* in a cereal field by strip-management. *Entomol. exp. appl.* 71: 111-20.

Zhao, Z. & McMurtry, J.A. 1990. Development and reproduction of three *Euseius* (Acari: Phytoseiidae) species in the presence and absence of supplementary foods. *Entomol. exp. appl.* 8: 233-42.

Zöllner, U. & Poehling, H.M. 1994. Influence of different aphid species on the efficiency of gall midge larvae (*Aphidoletes aphidimyza*) (Rond.) (Diptera, Cecidomyiidae). *Meded. Fac. Landbouwwet. Rijkuniv. Gent.* 59(2a): 281-86.

ENHANCEMENT

Overwintering, oviposition and population dynamics of hoverflies (Diptera: Syrphidae) in Northern Germany in relation to small and large-scale landscape structure

U. Krause[1] *& H.M. Poehling*[2]

[1]Institute for Plant Pathology, University of Göttingen,
Grisebachstr. 6, D-37077 Göttingen
[2]Institute for Plant Pathology, University of Hannover,
Herrenhäuser Str. 2, D-30419 Hannover

Abstract

A three year research project was started in 1992 on two differently structured sites in Northern Germany to find out whether an autochthonous overwintering syrphid population exists. Another aim was to examine the influence of field margins (hedges) on oviposition in spring and on population dynamics.

The large-scale experiments were carried out in winter cereals in Göttingen and Hiddestorf (near Hannover). The Göttingen site can be characterized as more diversified in comparison to Hiddestorf. One half of the plots at each site were close to field boundaries (hedges) and the other half was distant from hedges so that in addition to large-scale influences it should be possible to discuss small-scale effects of margins on the syrphid population separately for Göttingen and Hiddestorf.

Different sampling-techniques were established to measure the activity-density of hoverflies (Malaise traps, yellow trays), oviposition in spring (bait-plants) and to collect overwintering individuals (Malaise traps, yellow trays, sweep-netting).

Episyrphus balteatus, *Metasyrphus corollae*, *Sphaerophoria spp.*, *Syrphus ribesii* and *Syrphus vitripennis* occurred regularly in early spring and deposited their eggs on the bait-plants. These species were also collected in late autumn by sweep-netting so that overwintering can be supposed.

The females preferred more diversified sites for oviposition and at least in spring higher activity densities were observed in the more diversified locations close to hedges.

It is assumed that in more diversified areas the potential of biological aphid control by hoverflies is increased compared to landscapes with smaller proportions of field boundaries.

Arthropod natural enemies in arable land · II *Survival, reproduction and enhancement*
C.J.H. Booij & L.J.M.F. den Nijs (eds.). *Acta Jutlandica* vol. 71:2 1996, pp. 157-169
 ISBN 87 7288 672 2

Key words: syrphids, oviposition, overwintering, hedges, cereals, Malaise trap, yellow tray, bait-plant, activity density

Introduction

Aphids are important pests in crops causing heavy losses in yield and quality in some years. It is not always necessary to reduce them by applying insecticides because there is a range of polyphagous predators (e.g. spiders, carabids, staphylinids) and specialized antagonists like ladybirds, parasitoids and hoverflies. Especially the aphidophagous larvae of a number of hoverfly species are efficient predators of aphids in different crops due to their high predatory potential. Most adult hoverflies of about 400 species recorded in Germany feed on nectar and pollen, which is required mainly by females for the maturation of their reproductive system. The females being attracted by honeydew are able to find colonies of aphids very well and they oviposit close to them (Schneider 1948, Budenberg & Powell 1992). One key factor determining their efficiency is the temporal and spatial synchronisation between syrphids and aphids which shows a high degree of variability between years and even neighbouring regions (Ohnesorge & Schier 1989, Poehling et al. 1991).

The experiments focused on two main questions. Firstly, it should be examined whether an autochthonous, overwintering syrphid population exists in Northern Germany and if the development of an overwintering population later results in those high quantities of hoverflies noticed every year in July and August. Some authors (Gatter & Schmid 1990) assume a seasonal migration of syrphids in spring from the south to the north and vice versa in autumn to find a favourable place to overwinter as a possible reason for this phenomenon.

Another aim of research was to find out whether diversification of landscape has any influence on oviposition in spring and on population dynamics. It was assumed that the availability of overwintering habitats and food supply for adults (pollen) and for larvae (aphids) in early spring may be restricted in less diversified areas and may limit the build-up of autochthonous populations in time.

Research done until now has led to the result that managing field margins and vegetation strips within fields by providing attractive floral resources can increase local densities and diversity of hoverflies (Molthan & Ruppert 1988, Harwood et al. 1992, Ruppert 1993). Strips and boundaries are also used as refuges during harvest and as overwintering habitats (Nentwig 1989).
But an evaluation of field margins and strips has been restricted mostly to

regional research which seems to be problematic due to the high mobility of syrphids (Macleod 1992).

Methods

The experiments were carried out at landscape-scale in two regions of Northern Germany, one close to Göttingen and the other one near Hannover in Hiddestorf. The landscape in Hiddestorf can be characterized as less diversified with few hedges as a result of land consolidation in comparison to the landscape in Göttingen.

In the following, "boundaries" refers to hedges. The ecological evaluation of the hedges (e.g. density of margins/ha, botanical diversity, amount of indigenous shrubs and trees) has not been analysed yet so details about that are left aside now. In addition to the landscape level on both sites, differently sized fields were selected in the vicinity of hedges and distant from them. Each year new fields were chosen to carry out the experiments. Table 1 shows the number and size of the plots within the fields and the crops.

In each year one Malaise trap per plot was established for the measurement of activity density. In order to assess the oviposition of hoverflies in spring aphid infested bait-plants (winter wheat) were exposed in two plots. Ten pots with about five plants each were exposed between April and June on a field with a boundary and without a boundary in Göttingen as well as in Hiddestorf. Syrphid eggs deposited on the plants were reared in the laboratory until adult hoverflies emerged. The bait-plants were replaced by new ones twice a week.

During the winter of 1992/1993 one Malaise trap and yellow trays rested on a fallow in Göttingen to collect active and overwintering syrphids. In order to catch also inactive syrphids from the vegetation selective sweep-netting (i.e. not standardized) was carried out in autumn 1993.

Table 1. Number of plots of the experimental groups, plot size and crops.

	Göttingen		Hiddestorf		
	plots in fields with boundary	plots in fields without boundary	plots in fields with boundary	plots in fields without boundary	size of plots
1992	1 fallow	1 winter wheat	1 winter wheat	1 winter wheat	24 x 50 m
1993	4 winter wheat	4 winter wheat	3 winter wheat, 1 winter barley	3 winter wheat, 1 winter barley	24 x 70 m
1994	4 winter wheat	4 winter wheat	4 winter wheat	4 winter wheat	24 x 70 m

Results

Overwintering

Trapping results from malaise traps and yellow trays operating from late autumn to early spring, show that several species overwinter (Table 2).

The Malaise trap collected less species and individuals than the yellow trays. Most syrphids caught by the trays in autumn belonged to the species *E. balteatus* and in spring to *Melangyna* species. However, some more species (*M. corollae*, *S. vitripennis, Sphaerophoria spp.*) were collected in late autumn which were also caught by sweep-netting one year later in autumn (28.10.-4.11.93). The following individuals were caught in different habitats near Göttingen by sweep-netting: four males and eleven females of *E. balteatus*, two females of *Sphaerophoria spp.*, one male and two females of *M. corollae*, one female of *S. vitripennis* and one female of *S. torvus*. Many more females than males were caught confirming the results from the preceding year.

Table 2. Syrphid species with aphidophagous larvae, collected during autumn/winter 1992/1993 in a Malaise trap and yellow trays in Göttingen on a fallow, m=males, f=females.

	2.10	23.10	6.11	20.11	4.12-12.2	13.3	26.3	9.4	23.4
malaise trap	1m	0	0	0	0	0	0	0	0
Episyrphus balteatus	0	0	0			2 m 1 f	1 m	0	0
Melangyna spp.									
yellow trays									
Episyrphus balteatus	1 f	8 m 17 f	3 m 3 f	3 m 4 f	0	1 m	0	0	0
Metasyrphus corollae	0	2 f	0	0	0	0	0	0	0
Syrphus vitripennis	0	1 m 1 f	2 f	0	0	0	0	0	0
Melangyna spp.	0	0	0	0	0	36 m 14 f	1 m	13 m 25 f	1 m 7 f
Sphaerophoria spp.	1 m	0	0	0	0	0	0	0	0
Metasyrphus mellinum	0	1 m 3 f	0	0	0	0	0	0	0
B. elongata	0	1 f	0	0	0	0	0	0	0
P.punctulatus	0	0	0	0	0	0	0	0	2 m 1 f

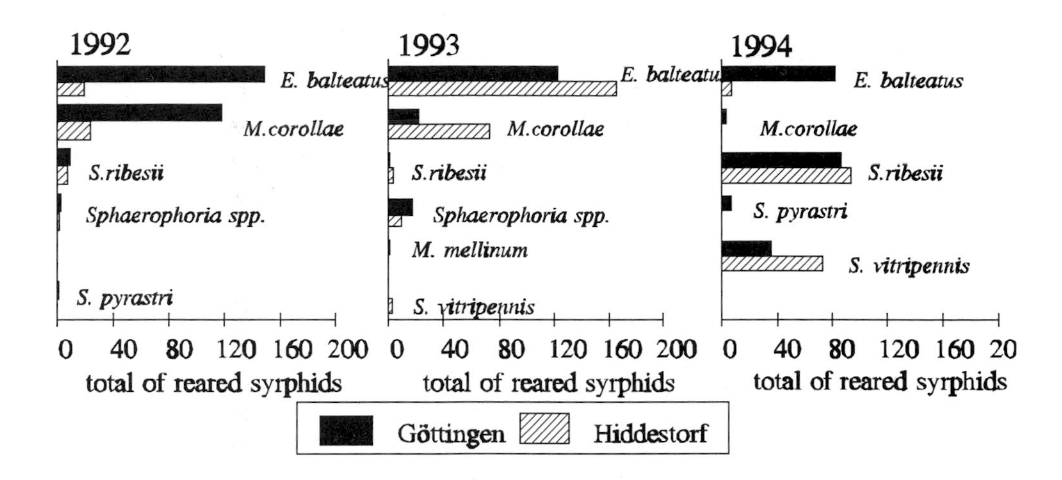

Fig. 1. Species composition of eggs laid by syrphids on bait-plants in 1992, 1993 and 1994; results of rearing in the laboratory.

Oviposition on bait-plants

Figure 1 describes the species composition of syrphids which laid eggs on bait-plants summed up for the different years and sites. Because of a similar species composition on plots with a boundary and without a boundary the adults of both experimental groups, which were reared in the laboratory, are combined. The bait plant trial showed that most species occurred every year but dominance structure varied.

The proportions of the species were very variable between the years: *M. corollae* was abundant in 1992 and 1993 but occurred scarcely in 1994. However, the two *Syrphus* species (S. *ribesii* and S. *vitripennis*) frequently deposited eggs on the bait-plants in 1994 whereas in the preceding years *S. ribesii* occurred rarely and *S. vitripennis* did not appear at all in 1992. The species which oviposit already in April and May are of special interest as far as the overwintering question is concerned. In 1992, at Göttingen and Hiddestorf the first oviposition occurred at the beginning of May with higher numbers in Göttingen than in Hiddestorf (Fig 2A). In most cases the oviposition on bait-plants exposed to fields with a boundary was higher than on those without boundaries. Oviposition already decreased at the end of May and remained at a very low level until the survey finished.

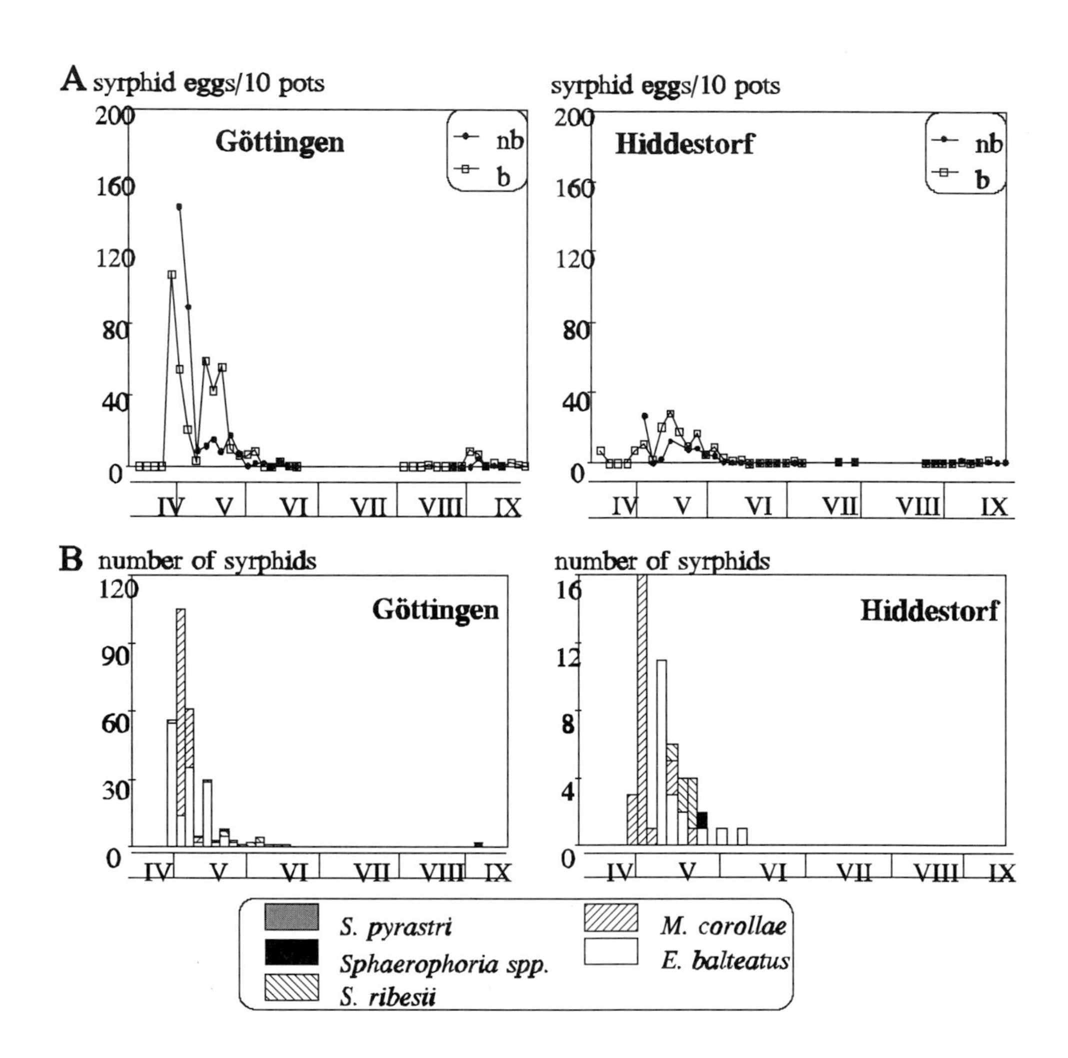

Fig. 2. Oviposition in Göttingen and Hiddestorf in 1992 (A) on plots with a boundary (b) and without a boundary (nb) and species composition (B).

Figure 2B represents the species composition of hoverflies which were reared in the laboratory from eggs deposited on the bait-plants. In Göttingen and Hiddestorf the species *E. balteatus* and *M. corollae* mainly contributed to oviposition in 1992.

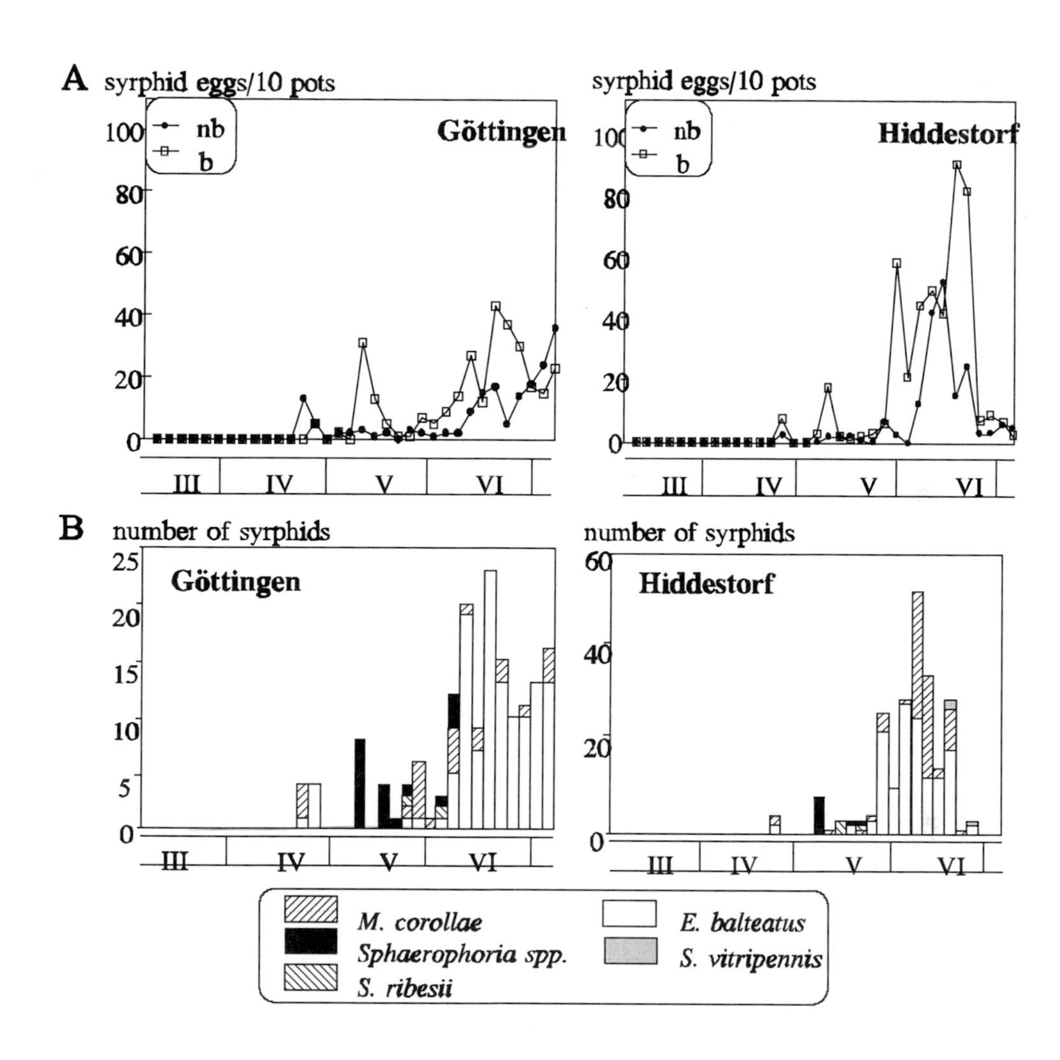

Fig. 3. Oviposition in Göttingen and Hiddestorf in 1993 (A) on plots with a boundary (b) and without a boundary (nb) and species composition of eggs reared (B).

As already observed in 1992 it turned out that in 1993 with few exceptions oviposition was higher on plots with boundaries in Göttingen and Hiddestorf (fig. 3A and 3B). The first eggs were deposited on the plots of both groups at the end of April and the oviposition had a peak in May at each site. The result indicates that fertile females cover great distances to find suitable places for

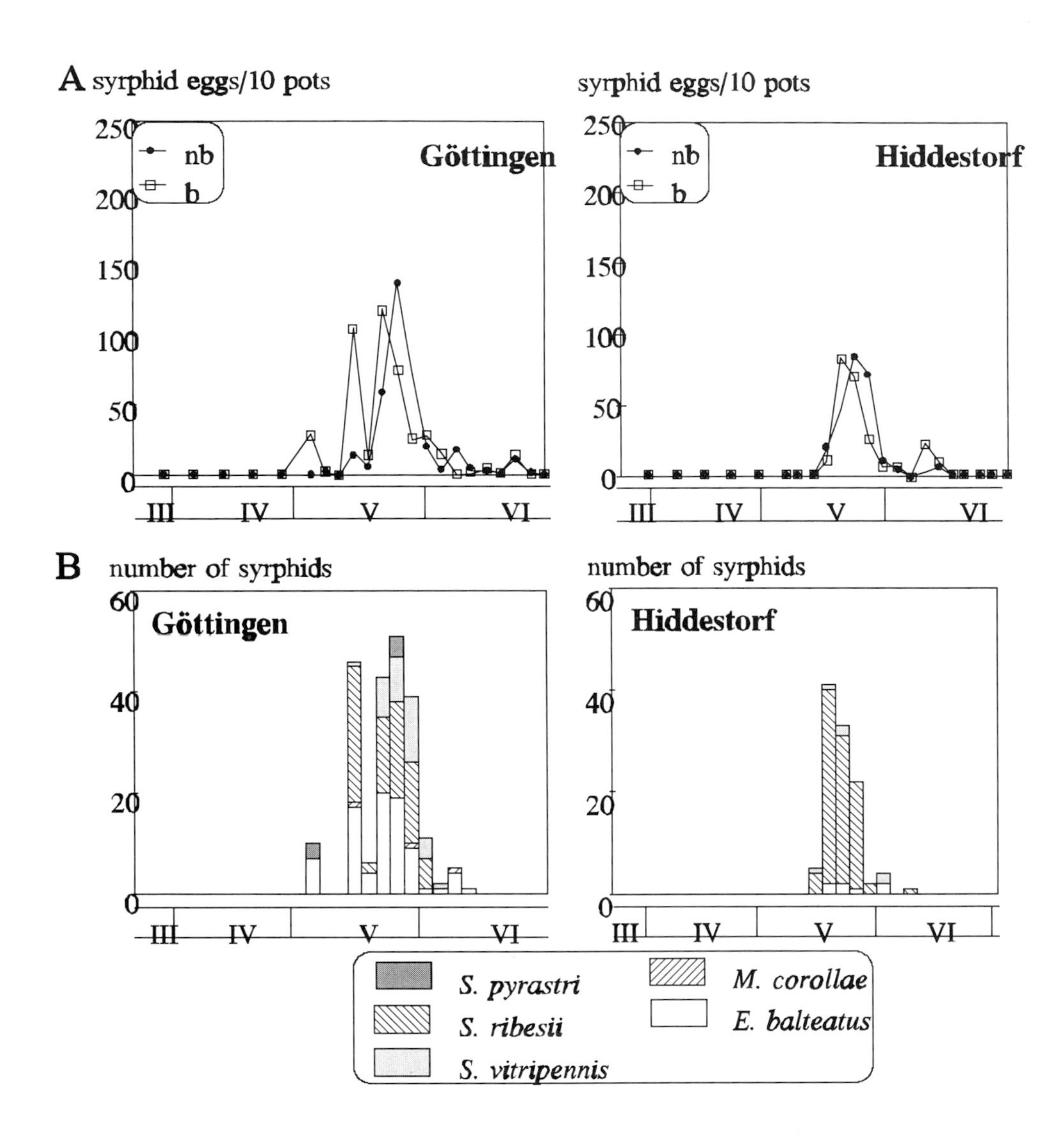

Fig. 4. Oviposition in Göttingen and Hiddestorf in 1994 (A) on plots with a boundary (b) and without a boundary (nb) and species composition of eggs reared (B).

oviposition represented by the fields with a boundary in this case. In comparison to the preceding year an increased oviposition was recorded on both sites in June.

This year, apart from *E. balteatus* and *M. corollae,* the eggs of *Sphaerophoria spp.* were found in higher numbers on the bait-plants (Fig. 3B).

As in the preceding years in 1994 oviposition began (Fig. 4A) at the end of April in Göttingen whereas in Hiddestorf it did not start until the middle of May with a maximum at the end of May at both sites. The period of oviposition covered more time in Göttingen than in Hiddestorf. In June oviposition decreased corresponding to the findings of 1992.

A clear difference can be noticed in comparison to the other years. In 1994 the syrphids did not show a distinct preference for oviposition on plots with a boundary. Also, egg laying was very different from the two preceding years. Oviposition of *E. balteatus* and *M. corollae* declined on both sites (fig. 4B) and the two *Syrphus* species occurred in higher numbers than in 1992 and 1993.

Table 3 shows the means of oviposition in the three years. A very clear trend of higher oviposition on plots with a boundary can be noticed. Only in 1994 was oviposition slightly increased in Hiddestorf on the plot without a boundary. In two out of three years (1992 and 1994) the total number of eggs was higher in Göttingen than in Hiddestorf, the less diversified experimental site.

Activity-density

In the following some more findings concerning activity densities assessed by Malaise traps in the years 1992, 1993 and 1994 are presented. Only species with aphidophagous larvae are included in the malaise trap data (Fig. 5-7), but these species represented the greatest part in the samples, varying between 92% and about 99% in the three years.

Table 3. Means of oviposition on bait-plants in 1992, 1993 and 1994 exposed to plots with a boundary (b) and without a boundary (nb) in Göttingen (Gö) and in Hiddestorf (Hi). Means followed by the same letter are not different ($P \leq 0.05$, Wilcoxon-test).

		1992		1993		1994	
		Gö	Hi	Gö	Hi	Gö	Hi
b	mean	17.0 a	4.5 a	8.5 ab	12.9 a	19.6 a	10.6 ab
	s.e.	28.0	7.5	12.2	23.8	34.4	22.6
nb	mean	13.6 ab	2.6 b	5.1 ab	5.3 b	16.0 ab	11.2 b
	s.e.	34.3	5.8	8.7	11.8	32.9	24.3
b+nb	mean	30.6 a	7.1 b	13.5 a	18.2 a	35.6 a	21.8 b
	s.e.	50.2	12.1	18.3	32.7	60.6	44.7

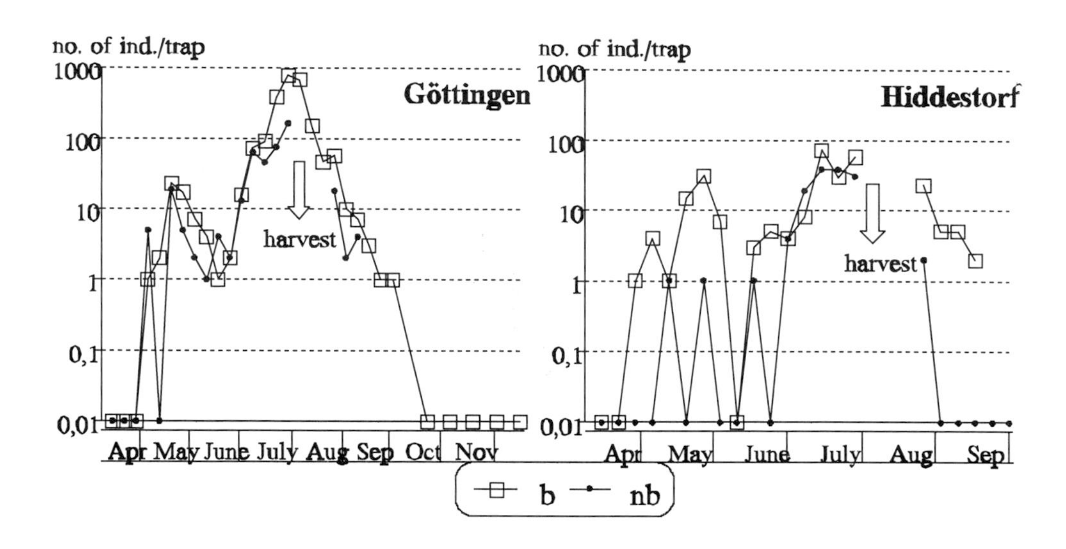

Fig. 5. Population dynamics assessed by Malaise traps in 1992.

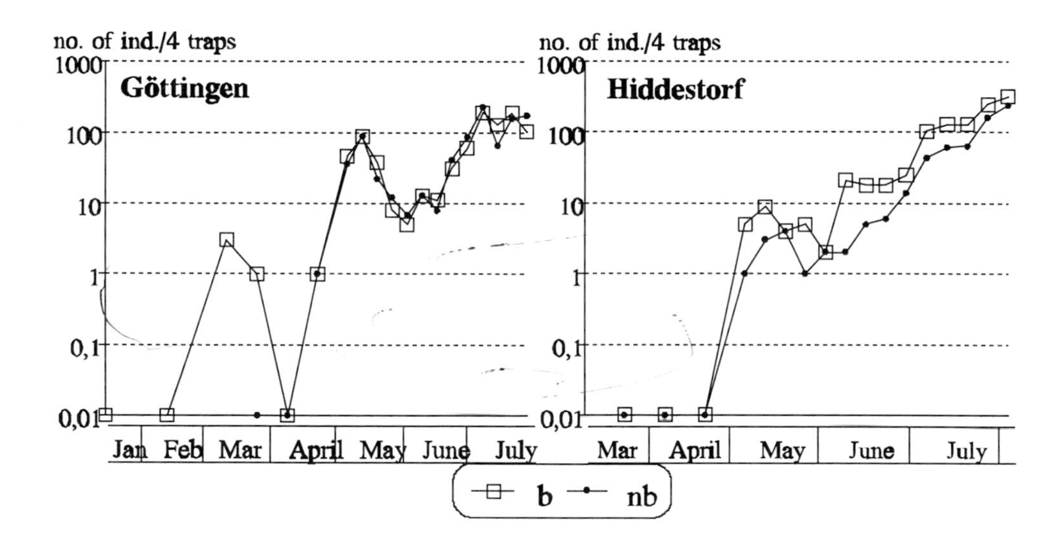

Fig. 6. Population dynamics assessed by Malaise traps in 1993.

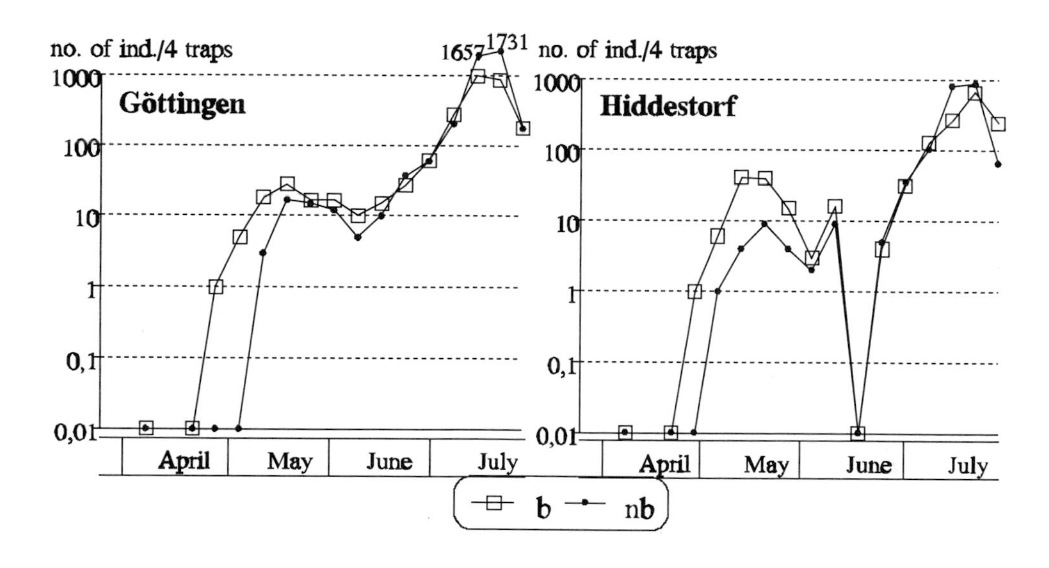

Fig. 7. Population dynamics assessed by Malaise traps in 1994.

In Göttingen and Hiddestorf in 1992 activity levels increased in May, a second peak occurred at the end of July on the fallow in Göttingen (Fig. 5). Because of harvest the traps had to be removed from the other winter wheat fields at that time, therefore we can only speculate upon the activity maximum on these plots. At least in spring the activity tended to be higher on plots with a boundary.

In 1993 the situation was similar to 1992 with an increasing activity density at the end of April in both experimental groups on both sites and a peak in July (Fig. 6). Only in Hiddestorf higher activity densities were found on plots with a boundary.

As in the two preceding years the first syrphids were collected in May giving higher numbers on plots with a boundary at that time (Fig. 7). In the middle of July the number of syrphids reached a maximum, but in contrast to the other years, the captures on plots without a boundary were higher in July compared to the other plots.

Discussion

In comparison to the malaise trap the yellow coloured trays attracted syrphids very well during autumn and winter 1992/1992. Therefore, they seem to be more appropriate especially in the flowerless season. The *Melangyna* species, mostly *M. quadrimaculata,* have only one generation per year which is active in spring. Therefore it is normal that so many individuals were collected in March and April. Species having several generations per year and which overwinter locally, should occur in autumn and spring, but this was only the case for *E. balteatus*. However, the results of the tray catches in autumn and winter 1992/1993 and of sweep-netting in 1993 indicate that several species are still active late in the season when a supposed movement southwards would have been stopped by that time. The number of females in the catches of the trays and of sweep-netting exceeded that of males, confirming the supposed overwintering of females (Schneider 1948). Another indication of a local hibernation is the early deposition of eggs on the bait-plants of those species collected in late autumn (i.e. *E. balteatus, M. corollae, S. vitripennis, Sphaerophoria spp.).* For these species it is very unlikely to assume a long-distance immigration at that time.

The species composition of the egg-laying syrphids on the bait-plants seems to vary more between the years than between the different sites which might indicate a climatic influence. The species which oviposited on the bait-plants very early in the year were always collected in late autumn in yellow trays and by sweep-netting. This is a clear indication that at least parts of the populations overwinter in Northern Germany.

In contrast to 1993 when egg-laying on the bait-plants was observed until the end of June, the oviposition in 1992 and 1994 decreased in May, probably due to the aphid infestations in these years. When overall aphid densities are very high, the aphid infested bait-plants are relatively unattractive. Therefore, the probability that hoverflies prefer the bait-plants decreases and oviposition is then dispersed in the whole field. In 1993 aphid densities were very low in general, hence the aphid infested bait-plants were attractive even late in the season.

For all three years the results of the bait-plant experiments indicated a preference for oviposition on plots with boundaries. The malaise trap data indicate that at least in spring the activity density of hoverflies is increased in plots with boundaries compared with plots without boundaries. The maximum activity density was not always higher on the plots with boundaries, but the active syrphids in July are quite late to play an important role as aphid antagonists in cereals.

Acknowledgements

This research was supported by the German Research Foundation and the Association for Applied Ecology.

References

Budenberg, W.J. & Powell, W. 1992. The role of honeydew as an ovipositional stimulant for two species of syrphids. *Entomol. exp. appl.*: 64: 57-61.

Gatter, W. & Schmidt, U. 1990. Wanderungen der Schwebfliegen am Randecker Maar. *Spixiana Suppl.* 15.

Harwood, R.W.J., Wratten, S.D. & Nowakowski, M. 1992. The effect of managed field margins on hoverfly (Diptera: Syrphidae) distribution and within-field abundance. *BCPC-Pests and diseases*: 1033-37.

MacLeod, A. 1992. Alternative crops as floral resources for beneficial hoverflies (Diptera: Syrphidae). *BCPC-Pests and diseases*: 997-1002.

Molthan, J. & Ruppert, V. 1988. Zur Bedeutung blühender Wildkräuter in Feldrainen und Äckern für blütenbesuchende Nutzinsekten. *Mitt. biolog. Bundesanstalt f. Land-u. Forstw. Berlin-Dahlem*: 85-99.

Nentwig, W. 1989. Augmentation of beneficial arthropods by strip-management. II. Successional strips in a winter wheat field. *Zeitschr. f. Pfl. Krankheiten u. Pfl. Schutz* 96 (1): 89-99.

Ohnesorge, B. & Schier, A. 1989. Regional differences in population dynamics of cereal aphids and their bearing on short-term forecasting. *Med. Fac. Landbouww. Gent* 54/3a: 747-52.

Poehling, H.M., Tenhumberg, B. & Groeger, U. 1991. Different pattern of cereal aphid population dynamics in Northern (Hannover-Göttingen) and Southern areas of West Germany. *Bull. IOBC/WPRS* 14 (4): 1-12.

Ruppert, V. 1993. Einfluß blütenreicher Feldrandstrukturen auf die Dichte blütenbesuchender Nutzinsekten insbesondere der Syrphinae (Diptera: Syrphidae). *Agrarökolgie* 8. Paul Haupt, Bern.

Schneider, F. 1948. Beitrag zur Kenntnis der Generationsverhältnisse und Diapause räuberischer Schwebfliegen (Syrphidae: Diptera). *Mitt. d. Schweiz. Entomol. Ges.* 21: 245-85.

Phacelia tanacetifolia flower strips: their effect on beneficial invertebrates and gamebird chick food in an integrated farming system

J.M. Holland & S.R. Thomas

The Game Conservancy Trust, Fordingbridge,
Hampshire, SP6 1EF, UK
Department of Biology, University of Southampton,
Bassett Crescent East,
Southampton, SO16 7PX, UK

Abstract
The impact of *Phacelia tanacetifolia* field margins on populations of Syrphidae, polyphagous predators, parasitoids and gamebird chick food was investigated at one of the LINK Integrated Farming Systems (IFS) sites in southern England. *Phacelia* strips were sown along the edge of winter wheat plots farmed using an integrated system and comparisons were made to the paired conventional farming system plots. *Phacelia* was attractive to Syrphidae and cereal aphid parasitoids, and populations in the adjacent winter wheat IFS plot were enhanced to some extent, but this was not sufficient to influence cereal aphid population development. Carabidae, Staphylinidae and Linyphiidae favoured the wheat crop in preference to the *Phacelia* strip. Gamebird chick food insects were more abundant in the *Phacelia* strip, primarily due to high densities of Miridae.

Key words: LINK Integrated Farming Systems, field margin, Syrphidae, Carabidae, Staphylinidae, Linyphiidae, Miridae

Introduction

An option for arable crop production currently being investigated throughout Europe is the adoption of integrated farming methods (Holland et al. 1994). One of the main objectives of integrated farming is to utilise natural regula-

Arthropod natural enemies in arable land · II Survival, reproduction and enhancement
C.J.H. Booij & L.J.M.F. den Nijs (eds.). *Acta Jutlandica* vol. 71:2 1996, pp. 171-182
 ISBN 87 7288 672 2

tory mechanisms and thereby reduce the need for inputs of pesticides, fertilizers and fuels (Jordan and Hutcheon 1994). The establishment of a flower border strip which is attractive to cereal aphid predators, for example hoverflies (Diptera: Syrphidae), is one technique which may enhance cereal aphid control and so reduce the need for and frequency of aphicides. Adult hoverflies require proteins from pollen to mature their reproductive systems (Schreiber 1948). Eggs are then laid in association with aphid colonies on which the larvae subsequently feed (Dean 1974). Many hymenopteran parasitoids also feed on non-host flowers to obtain nectar or pollen (Jervis et al. 1993) which can increase fecundity and longevity (Jervis & Kidd 1986). *Phacelia tanacetifolia* and other flowering plants have been shown in selection tests to be attractive to hoverflies (Lovei *et al.* 1993) and hymenopteran parasitoids (Jervis et al. 1993). Whether *Phacelia tanacetifolia* flower borders also harbour polyphagous predatory invertebrates and gamebird chick food insects which are associated with field margins has not been previously investigated.

A new programme of research was started in 1992 to investigate integrated farming – the LINK Integrated Farming Systems project (Holland et al. 1994b). As part of a programme of measures to reduce the inputs of agrochemicals, strips of *Phacelia* were established around the edge of cereal fields at some of the experimental sites. Their impact on hoverfly, parasitoids, polyphagous predators and chick food prey items was evaluated at one of these sites.

Materials and Methods

At the LINK IFS site in north Hampshire the conventional farm practice (CFP) and integrated farming system (IFS) were compared using adjacent pairs of plots (min. size 5 ha, min. width 100 m). Detailed descriptions of the experimental design are available in Prew (1993). In this study four pairs of winter wheat plots were used annually in 1993 and 1994. A 1 m wide strip of *Phacelia* was sown (1 g m^{-2}) on the 20th April 1993 and 6th April 1994 along the longest edge (300-400 m) of the four winter wheat IFS plots. No strip was sown in the CFP plots which were used as a comparison. To sample airborne invertebrate predators and parasitoids a transect of fluorescent yellow water traps (19 cm diameter) were located at crop height. These were positioned with one in the *Phacelia* strip and the remainder in a line at 45° and at distances of 10, 25, 50, 75 and 100 m to each *Phacelia* strip in the IFS plots. Another transect of traps was also located on the opposite side of the field but adjacent to the CFP plots which did not have a *Phacelia* strip. The traps were emptied weekly from the 13th June until the 16th July 1993 and from the 21st June until the 18th July 1994. *Phacelia* started to flower on the 21st June 1993 and on the 28th May 1994. The contents of each trap were stored separately in 70% alcohol until identification. The following taxa only are discussed: Syrphidae, *Aphidius* spp. (Braconidae), other Braconidae (mostly *Praon* spp. and *Toxares* spp.) and *Platypalpus* spp. (Diptera: Empididae). In addition the total number of cereal aphid parasitoids and their hyperparasitoids were identified in the 1994 samples. Numbers caught in each taxa on each sampling occasion were transformed (log x+1) and analysed separately using repeated measures Analysis of Variance (ANOVA) with farming system and distance from field margin as factors (SPSS Inc. 1993). Each year was analysed separately.

All CFP and IFS plots were sampled using a Dietrick vacuum insect sampler (D-vac) on the 21st July 1993 and on the 23rd June 1994. Two transects of five samples (10 s suction, repeated five times at 2 m intervals) were taken in the centre of each plot starting 30 m from the common boundary and progressing into the plot. Five D-vac samples were taken in the IFS plots on 7th July 1993 and 21st June 1994 (10 s suction, repeated five times at 2 m intervals) from the *Phacelia* strip, at 3 m from the crop edge (in the selectively sprayed Conservation Headland area) and at 10 m from the crop edge. All samples were frozen and the invertebrates identified as for the water trap samples in addition to all Carabidae, Staphylinidae, Linyphiidae were identified. In the 1994 samples also the total number of cereal aphid parasitoids, their hyperparasitoids, all other parasitoids and the total gamebird chick food prey items (Carabidae, Staphylinidae, Chrysomelidae, Curculionidae, Miridae, Nabidae, Hymenoptera larvae and Lepidoptera larvae) were identified. To determine differences between sampling positions individual taxa were analysed using one-way ANOVA.

Cereal aphid numbers, species, life-stage (adult or nymph) and aphid mummies were assessed on 50 tillers in 1993 and 25 tillers in 1994 for each winter wheat plot on five occasions. The total number of grain aphids (*Sitobion avenae* F.) per 50 or 25 tillers (log x+1) in each farming system on the five sampling occasions were compared for each year using repeated measures ANOVA. The analysis was repeated for rose-grain aphids (*Metopolophium dirhodum* Wlk.) and cereal aphid mummies in each year. To determine whether the population age structure varied between the two farming systems the proportion of nymphs was calculated.

Results

1993

The numbers of *Aphidius* spp., other Braconidae, Syrphidae and *Platypalpus* spp. in the water traps only differed significantly (P<0.05) with time of sampling (Table 1). There was a trend towards more Syrphidae in the *Phacelia* (0 m) and up to 25 m into the crop and towards more Braconidae in

Table 1. Total number per water trap at each distance from the field margin and the pooled standard error (SE) for each invertebrate group captured in the four plots with *Phacelia* (IFS) and four without (CFP) during 1993.

distance from margin	Aphidius spp.		Other braconidae		Syrphidae		Platypalpus spp.	
	CFP	IFS	CFP	IFS	CFP	IFS	CFP	IFS
0 m	14.5	16.5	56.8	38.8	16.5	28.5	116	251
10 m	17.0	15.0	50.0	73.3	17.0	22.3	249	318
25 m	15.8	18.0	62.5	112.3	16.0	25.5	317	408
50 m	14.3	18.5	39.8	101.5	15.3	17.0	235	453
75 m	18.3	18.8	80.8	96.0	14.3	20.0	288	397
100 m	17.3	14.3	66.8	44.5	14.5	15.3	301	444
SE	3.2		25.3		5.5		63.7	

Table 2 Mean number per 0.5 m^2 (±SE) of each invertebrate group in the field centre and margin D-vac samples in 1993.

	Plot centre		Field margin		
Invertebrate group	CFP	IFS	*Phacelia*	Crop edge	10m into crop
Carabidae	2.0 (± 1.3)	1.3 (± 0.9)	0.2 (± 0.2)	0.1 (± 0.1)	0.5 (± 0.2)
Staphylinidae	4.6 (± 2.7)	3.1 (± 1.5)	0.9 (± 0.4)	1.0 (± 0.4)	1.5 (± 0.3)
Linyphiidae	5.6 (± 1.3)	7.1 (± 1.3)	4.4 (± 0.8)	7.4 (± 1.3)	9.6 (± 1.0)
Aphidius	11.9 (± 2.0)	10.8 (± 1.6)	10.6 (± 2.4)	5.2 (± 0.7)	7.6 (± 0.9)
Other Braconidae	8.3 (± 0.8)	9.6 (± 0.3)	186.9 (± 80.1)	26.7 (± 6.1)	16.5 (± 2.2)
Syrphidae	0.0 (± 0.0)	0.0 (± 0.0)	1.9 (± 0.6)	0.6 (± 0.3)	0.4 (± 0.2)
Platypalpus spp.	11.9 (±1.0)	15.8 (± 2.8)	3.0 (± 0.9)	6.3 (± 1.6)	8.9 (± 1.1)
Chick food insects	6.7 (± 4.0)	4.5 (± 2.2)	34.6 (± 10.4)	8.5 (± 2.9)	3.6 (± 0.8)

the IFS plots. The predatory flies (*Platypalpus* spp.) were the most abundant predatory species in the wheat crop, especially in the IFS plots.

There were no significant differences between the numbers of the taxa identified in the IFS and CFP mid-field D-vac samples (Table 2). Neither was there a significant difference between the field margin D-vac suction sample positions. This was expected because of the low between-subjects degrees of degrees of freedom. The density of other Braconidae was very high in the *Phacelia* strip and this was also found to a lesser extent for *Aphidius* spp., Syrphidae and chick food insects (Table 2). The polyphagous invertebrates (Carabidae, Staphylinidae and Linyphiidae) appeared less abundant in the *Phacelia* strip compared to the adjacent wheat crop.

There was no significant difference in the number of grain or rose-grain aphids between the CFP and IFS plots or between the different sample dates and no interaction effect (Fig. 1). Neither was there a significant difference in the proportion of nymphs of either species between the two farming systems

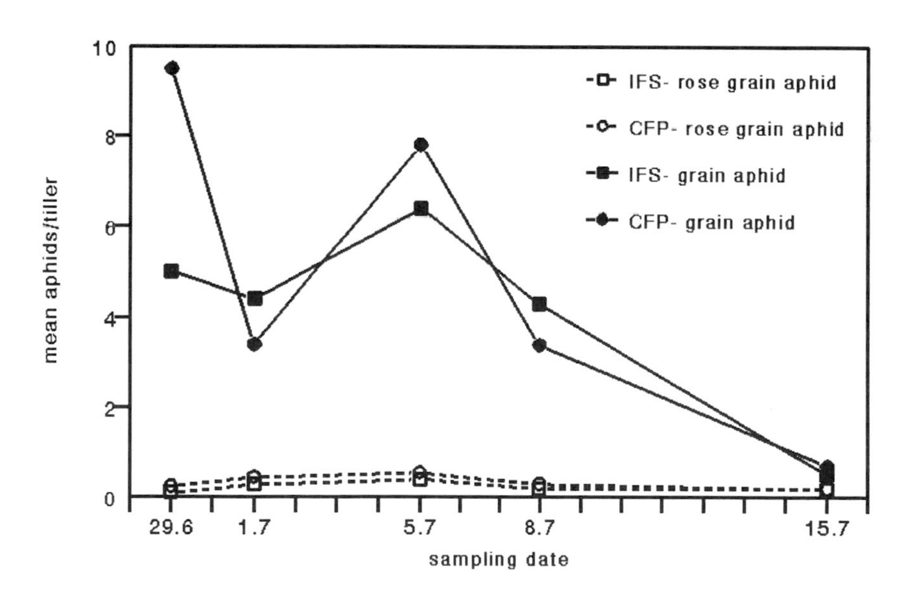

Fig. 1. Mean number of grain aphids and rose-grain aphids per tiller in CFP and IFS plots during 1993.

Table 5. Mean number per 0.5 m^2 (± SE) of each invertebrate group in the field centre and margin per D-vac samples in 1994

	plot centre		field margin		
invertebrate group	CFP	IFS	*Phacelia*	crop edge	10 m into crop
Carabidae	0.4 (± 0.3)	0.3 (± 0.1)	0.0 (± 0.0)	0.1 (± 0.0)	0.2 (± 0.1)
Staphylinidae	1.1 (± 0.7)	0.9 (± 0.7)	0.4 (± 0.1)	0.8 (± 0.3)	0.8 (± 0.3)
Linyphiidae	3.4 (± 0.5)	3.2 (± 0.3)	3.2 (± 1.3)	5.0 (± 0.6)	4.0 (± 0.5)
Aphidius spp.	12.2 (± 0.9)	11.9 (± 0.5)	12.8 (± 2.6)	6.1 (± 0.9)	5.2 (± 0.9)
Other Braconidae	11.3 (± 0.8)	10.9 (± 0.5)	5.6 (± 3.6)	0.6 (± 0.2)	6.5 (± 2.2)
Total cereal parasitoids	5.5 (± 1.6)	4.8 (± 1.0)	33.0 (± 11.3)	8.5 (± 1.3)	8.0 (± 1.4)
Total cereal hyperparasitoids	1.7 (± 0.7)	0.7 (± 0.3)	0.8 (± 0.3)	0.5 (± 0.1)	0.4 (± 0.2)
Syrphidae	0.0 (± 0.0)	0.0 (± 0.0)	0.5 (± 0.3)	0.0 (± 0.0)	0.0 (± 0.0)
Platypalpus spp.	7.1 (± 2.0)	6.6 (± 2.1)	3.4 (± 1.2)	8.4 (± 1.9)	6.0 (± 1.3)
Chick food insects	3.2 (± 1.3)	2.0 (± 0.8)	25.4 (± 5.8)	2.7 (± 0.7)	6.3 (± 1.0)

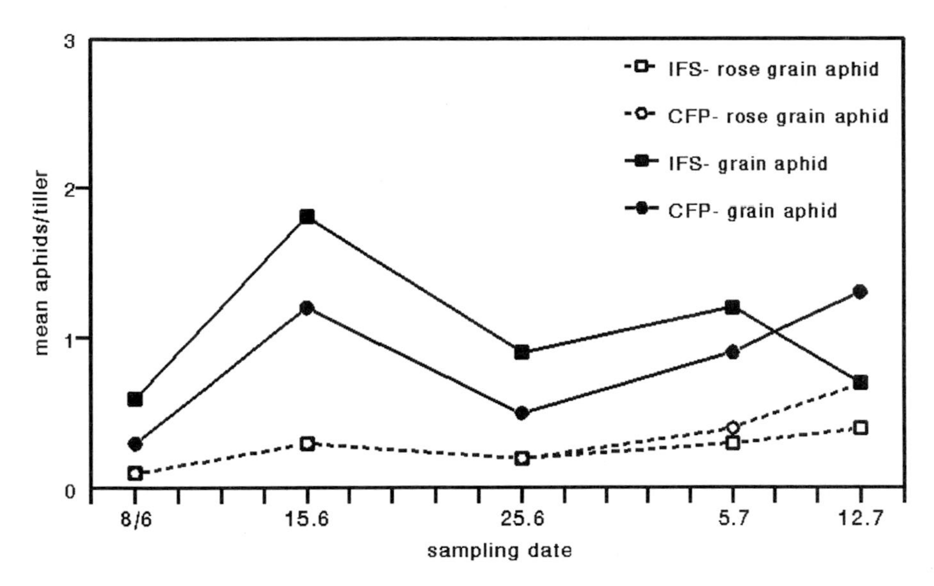

Fig. 2. Mean number of grain aphids and rose-grain aphids per tiller in CFP and IFS plots during 1994.

fered significantly (P<0.05) between the two farming systems with a greater decline in the IFS plots (Fig. 2). No significant effects were found for rose-grain aphids or for the proportion of nymphs of both species (Table 3). The number of mummified aphids was low (approximately 0.2 per tiller) and was not significantly different between the two farming systems.

Discussion

The results of the D-vac suction samples and water traps although generally not conclusive did indicate some possible trends. Syrphidae were utilising the pollen source provided by *Phacelia* as found by Harwood et al. (1994) and were subsequently distributing into the adjacent crop. However the *Phacelia* strip only increased numbers of Syrphidae into the crop for up to 25 m in 1993 and only 10 m in 1994. Studies in New Zealand also found Syrphidae only dispersed up to 15 m into the crop from *Phacelia* strips (Lövei et al. 1993). Gut dissections of all the Syrphidae captured in water traps in 1993 revealed that a higher proportion of Syrphidae captured in the plot with *Phacelia* contained *Phacelia* pollen. The provision of extra pollen did not, however, increase egg production because no difference was found in the proportion with eggs (Holland et al. 1994). The impact of providing such a massive pollen source on aphid-specific predators like Syrphidae requires further investigation to determine whether populations of Syrphidae are increasing or redistributing, so lowering levels of aphid control in other areas. In addition, because Syrphidae are capable of moving large distances, the use of *Phacelia* as a green cover on set-aside land may act as a sink thus influencing Syrphidae distribution in the arable landscape.

The Braconidae (mostly *Praon* spp. and *Toxares* spp.) and to a lesser extent *Aphidius* spp. were more numerous in the D-vac samples taken in the *Phacelia* in 1993 and 1994 and the total density of cereal parasitoids was also higher in the *Phacelia* in 1994. No aphids were found in the *Phacelia* strip which would attract cereal aphid parasitoids, therefore, it was either the *Phacelia* flowers which were attractive or the cover which was provided. Preliminary wind tunnel investigations indicated that *Aphidius* spp. were positively attracted to *Phacelia* flowers (pers. comm. Courts 1993). The cereal parasitoids were not, however, more numerous in the water traps positioned in the *Phacelia* indicating activity was low and an influx of cereal parasitoids into these traps was not found when the started to flower. In contrast, there were more other Braconidae and thus total cereal parasitoids in the water traps positioned in the crop in the IFS plots during 1993 and 1994, but this was not reflected in the D-vac samples. This indicated that activity of cereal parasitoids was higher in the IFS plots because the capture rate of water traps is strongly dependant on activity, but their density as indicated by the D-vac sampling was similar. The capture of more other Braconidae for both farming systems in the water traps in 1993, when aphid numbers were low compared to 1994, confirmed that water trap capture rate was influenced by activity. There was no increase in the proportion of parasitised cereal aphids in the IFS plots in 1993 or 1994 as a result of higher cereal parasitoid activity. In conclusion, *Phacelia* was attractive to some cereal parasitoid species and although did not appear to increase their density, there was an increase in activity in the IFS plots which contained the *Phacelia* strip. Whether feeding on *Phacelia* flowers increases activity possibly as a result of improved reproductive capacity or energy requires further investigation.

The predatory flies (*Platypalpus* spp.) were more abundant in the water trap samples from the IFS plots in 1993, however, this was not found in 1994. Empid flies are only caught in coloured traps when they are searching for food or egg laying sites (Baillot & Tréhin 1974) thus activity was higher in the IFS plots during 1993. The D-vac sampling indicated that there were no differences in density between the CFP and IFS plots in 1993 or 1994. Both the water traps and D-vac sampling showed that the predatory flies avoided the *Phacelia* strip, probably because of lack of prey.

Polyphagous invertebrate predator populations were lower in the *Phacelia* probably because prey was sparse. Introduced flora would be expected to host less plant-feeding invertebrate species. Sampling was, however, carried out during day time when many polyphagous predators are inactive (Vickerman & Sunderland 1975) and night sampling should be carried out. Chick food was more abundant in the *Phacelia* compared to the adjacent crop, although mirid bugs formed the greater proportion of the catch and diversity was low. This has implications for farmland birds where *Phacelia* is sown on set-aside land.

In summary, although *Phacelia* was attractive to some beneficial insects, notably Syrphidae and some cereal parasitoid species, their numbers were not increased to any great extent within the crop and aphid control was not enhanced. Incorporating native flowering species into the *Phacelia* may encourage a more diverse range of beneficial species. Umbelliferae would be suitable as these are attractive to a wide range of parasitoid species (Jervis et al. 1993) and Syrphidae. Alternatively allowing some broad-leaved weeds to flower within the crop may provide sufficient floral resources for the Syrphidae and cereal parasitoids, whilst also providing cover for ground active species and encouraging greater dispersion within the crop.

Acknowledgements

The authors thank the Manydown Company for allowing access to their sites. The study formed part of the LINK Integrated Farming Systems project which is jointly funded by the Ministry of Agriculture, Fisheries and Food, Scottish Office Agriculture and Fisheries Department, Zeneca Agrochemicals, Home Grown Cereals Authority (Cereals and Oilseeds) and British Agrochemicals Association.

References

Baillot, S. & Tréhin, P. 1974. Variations de l'attractivité des pièges colorés de Moericke en function de la localisation spation-temporelle de l'émergence, des comportements sexuels et des phases de dispersion de quelques espèces de Diptères. *Ann. Zool. - Ecol. Anim.* 6: 575-84.

Dean, G.J. 1974. Effects of parasites and predators on the cereal aphids *Metopolophium dirhodum* and *Macrosiphum avenae* (Hemiptera: Aphididae). *Bull. Entomol. Res.* 20: 209-24.

Harwood, R.W.J., Hickman, J.M., MacLeod, A., Sherratt, T.N. & Wratten, S.D. 1994. Managing field margins for hoverflies. In: *Field Margins: Integrating Agriculture and Conservation.* BCPC Monograph No. 58, 147-52.

Holland, J.M., Frampton, G.K., Wratten, S.D. & Cilgi, T. 1994 Arable acronyms analyzed – a review of integrated farming systems research in Western Europe. *Ann. of Appl. Biol.* 125: 399-438.

Holland, J.M., Thomas, S.R. & Courts, S. 1994. *Phacelia tanacetifolia* flower strips as a component of integrated farming. In: Boatman, N.D. (ed.) *Field Margins: Integrating Agriculture and Conservation. BCPC Monograph No. 58*, Farnham: BCPC, 215-20.

Jervis, M.A. & Kidd, N.A.C. 1986. Host-feeding strategies in hymenopteran parasitoids. *Biol. Rev.* 61: 395-434.

Jervis, M.A., Kidd, N.A.C., Fitton, M.G., Huddleston, T. & Dawah, H.A. 1993. Flower-visiting by hymenopteran parasitoids. *J. Nat. Hist.* 27: 67-106.

Jordan, V.W.L. & Hutcheon, J.A. 1994. Economic viability of less-intensive farming systems designed to meet current and future policy requirements: 5-year summary of the LIFE project. *Asp. Appli. Biol. 40, Arable farming under CAP reform*, 61-68.

Lövei, G.L., Hickman, J.M., McDougal, D. & Wratten, S.D. 1993. Field penetration of benefical insects from habitat islands: hoverfly dispersal from flowering strips. *Proc. 46th New Zealand Plant Prot. Conf.*: 325-28.

Prew, R.D. 1993. Development of Integrated Farming Systems for the UK.*Proc. HGCA R & D Conf. 1993*, London: HGCA, 242-54.

Schreiber, F. 1948. Beitrag zur kenntnis der Generationsverhaltnisse and Diapause rauberischer Schwebfliegen. *Mitt. Scheiz. Entomol. Ges.*, 21: 249-85.

SPSS Inc. 1993. SPSS for Windows: Advanced Statistics, Release 6.0. SPSS Inc., Chicago, USA.

Vickerman, G.P. & Sunderland, K.D. 1975. Arthropods in cereal crops: nocturnal activity, vertical distribution and aphid predation. *J. Appl. Ecol.* 12: 755-65.

Management of arable crops and its effects on rove-beetles (Coleoptera: Staphylinidae), with special reference to the effects of insecticide treatments

Joachim Zimmermann & Wolfgang Büchs

Federal Biological Research Centre for Agriculture and Forestry,
Institute of Plant Protection in Field Crops and Grassland,
Messeweg 11/12, D-38104 Braunschweig, Germany

Abstract
Using emergence traps and pitfall traps, staphylinid beetles were investigated under the influence of different inputs of pesticides and fertilizers.

The life-cycles of the dominant species and their susceptibility to insecticides was investigated. Reproducing populations were most sensitive to insecticide applications. For spring treatments for example, spring breeders were the most affected, and autumn breeders were harmed mostly by autumn applications. On the other hand treatments during the time of the hatch of the imagines did not have significant effects on the fauna. The individuals are probably not exposed to the pesticides whilst living as larvae in the soil. After hatching they immediately enter a phase of dispersion, in which they leave the habitat, so avoiding contact with pesticides. To assess the effects of pesticides on rove beetles it is suggested that the initial reduction of abundance in addition to the emergence of the next generation (delayed effects) is measured.

Key words: life cycles, reproduction, *Atheta triangulum, A. pittionii, A. elongatula, A. aegra, A. palustris, Aloconota gregaria, Tachyporus spp., Philothus spp., Lesteva longelytrata, Liogluta alpestris nitidula, Omalium caesum, Oxypoda exoleta, O. longipes*

Introduction

Life-cycles of rove-beetles are mostly unknown. Studies by Topp (1979), Hartmann (1979) and Assing (1993) examined several species of woodlands,

Arthropod natural enemies in arable land · II *Survival, reproduction and enhancement*
C.J.H. Booij & L.J.M.F. den Nijs (eds.). *Acta Jutlandica* vol. 71:2 1996, pp. 183-194
 ISBN 87 7288 672 2

meadows and a dune. Life-cycles of single species were investigated by Lipkow (1966), Kasule (1968, 1970), Eghtedar (1970) and Schimke (1978). The coenosis of Staphylinidae in agroecosystems had not been examined till now in this context.

Information on Staphylinid life-cycles is necessary for assessing the effects of pesticide treatments in field trials. Sprick (1990) for example, tested the initial toxicity of pesticides at different parts of the year. Pesticides seem to be less effective during the hatching of the imagines and during the period when populations are living as larvae in the soil. During this period individuals are obviously protected from direct topical contamination.

In this experiment effects of spring, summer and autumn applications on the rove-beetle coenosis were measured, and it was elucidated which species were most sensitive to treatments at different times of the year.

Material and Methods

The trials were carried out on a field near Braunschweig, Lower Saxony in Nothern Germany from 1989 to 1992. The field was divided into four different cultivation intensities, which are defined as follows:

I_0 = Crop production without input of pesticides (except seed dressing). Minimum input of fertilizers
I_1 = Extensive crop production with a suboptimal input of nitrogen and low input of pesticides
I_2 = Integrated crop production with the aim of obtaining a high yield by minimizing the input of pesticides as far as possible
I_3 = Intensive crop production by utilization of all allowed agents to obtain a maximum yield. Prophylactic use of pesticides

The fauna was caught using six pitfall traps and five emergence traps per plot. Each emergence trap covered a soil surface area of 1 square meter. The location of each emergence trap was changed once a month, the traps were emptied twice a month.

The field was cultivated with a crop rotation of three years with sugar beet, winter wheat, winter barley and a catch crop. The soil type is löss soil and the whole region around the experimental site consists of intensively cultivated farmland, with a low proportion of non-crop habitats, for example hedges, field margins, pasture land or woodlands (further information is given by Büchs (1993, 1994).

Results

Life cycles of rove-beetles

To elucidate the little understood life-cycles of the rove-beetles, the abdomina of females were dissected to look for mature eggs in the gonads. The reproductive phase of a particular species was defined as the time when mature eggs were recorded. To determine the period of hatching of imagines, the appearance of immature (callow) specimens was registered.

In Table 1 the life-cycles of the most frequent species found in the trial are listed. The reproductive periods showed time related niches in the different seasons. Species reproducing very early in the year were mostly very small (*Atheta triangulum, Aloconota gregaria, Atheta pittionii, Atheta*

Table 1. Life-histories of dominant species and the most suitable sampling method. A= adult hibernation, L= larval hibernation, ET= emergence trap, PT= pitfall trap.

Species	Reproduction	Hatching of Imagines	Sampling method	Hibernation Type
Atheta triangulum KRAATZ	II-IV	VI	ET	A
Aloconota gregaria ERICHSON	II-IV	V-VI	ET/PT	A
Atheta pittionii SCHEERPELTZ	IV	VI-VII	ET	A
Anotylus rugosus FABRICIUS	IV	V-VI	ET/PT	A
Atheta elongatula GRAVENHORST	IV	VI	ET	A
Tachyporum hypnorum FABRICIUS	V-VI	VII-VIII	ET/PT	A
Philonthus cognatus STEPHENS	V-VI	VIII	PT	A
P. rotundicollis MÉNÉTRIÉS	V-VI	VIII	PT	A
Oxypoda exoleta ERICHSON	VIII	VI	ET/PT	L
Atheta aegra HEER	X-IV	V-VI	ET	A/L
Omalium caesum GRAVENHORST	X-XI	V-VI	ET/PT	A/L
Liogluta alpestris nitidula KRAATZ	X-XI	V-VI	ET (PT?)	A/L
Lesteva longelytrata GOEZE	XI-II	V	ET/PT	A/L

elongatula). The larger ones follow in (early) summer (*Tachyporus spp., Philonthus spp.*). Surprisingly, many species reproduce in autumn and winter *Atheta aegra, Lesteva longelytrata, Liogluta alpestris nitidula, Omalium caesum*). In contrast, the hatching of imagines of all species took place in a relatively short period of three to four months from May to August.

In the column "sampling method" (Table 1) those methods are listed by which a species can be caught in the two main periods of its adult life: reproduction and the hatching of imagines. Many rove-beetles show a high

Table 2. Insecticide treatments in 1989, 1990 and 1992 in various crops under different cultivation intensity.

Date	Cultivation Intensity		
	I_1	I_2	I_3
		sugar beet, 1989	
31-03-1989	Carbofuran-Gr	Carbofuran-Gr	Carbofuran-Gr
27/29-05		Pirimicarb	Pirimicarb
08/13-06	Pirimicarb	Pirimicarb	Demeton-S-Methylsulfoxid
22/23/29-06	Pirimicarb	Pirimicarb	Pirimicarb
06/10-07		Pirimicarb	Pirimicarb
		winter wheat, 1989/90	
04-05-1990		Fenvalerate + Pirimicarb	Fenvalerate + Pirimicarb
18-05		Deltamethrin	Deltamethrin + Lamda-Cyhalothrin
12/14-06		Pirimicarb	Parathion
21-06	Pirimicarb	Pirimicarb	Fenvalerate
		winter barley, 1990/91	
23-10-1990		Fenvalerate	Fenvalerate
		sugar beet, 1992	
22-04-1992		Carbofuran-Gr	Carbofuran-Gr
29-05		Pirimicarb	Deltamethrin
12-06		Pirimicarb	Deltamethrin

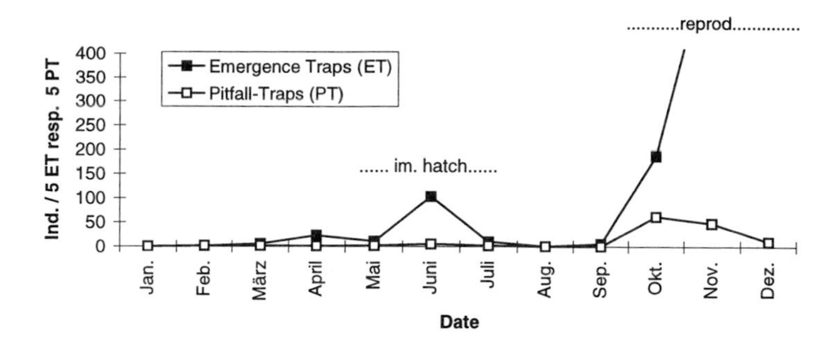

Fig. 1. *Atheta aegra*: Comparison of phenologies recorded with pitfall traps and emergence-traps in the cultivation intensity I_0 (winter wheat 1990). im.- hatch = hatching of imagines; reproduc. = period of reproduction.

positive phototactic orientation during the hatching of the imagines, in order to leave the fields, so that pitfall trapping is not a suitable method during this period. For example, in Fig. 1 the phenology of *Atheta aegra* in 1990 is shown for both methods. During reproduction, activity could be measured by both methods, but during the hatching of imagines by emergence traps only. On the other hand, pitfall traps seem to be the appropriate method for larger species of rove-beetles, which show greater activity on the soil surface (e.g. *Philonthus spp*.). These findings have to be taken into consideration when intending to measure effects of certain treatments in different seasons of the year.

Effects of insecticides

In 1989 and 1992 sugar beet was grown on the experimental site. In Table 2 the insecticide treatments are listed. In 1989 the strongest effects on the fauna occurred when granules of the insecticide carbofuran were applied during the reproductive period of the spring breeders. The type of effects can be subdivided into initial and delayed reduction of emergence. Initial effects are perceptible e.g. in the phenology of Anotylus rugosus (Fig. 2A). The application took place while the beetles showed epigeic activity, the males for copulation and the females for egg laying. Up to six weeks after the application more than 400% more specimens were found in the untreated plot

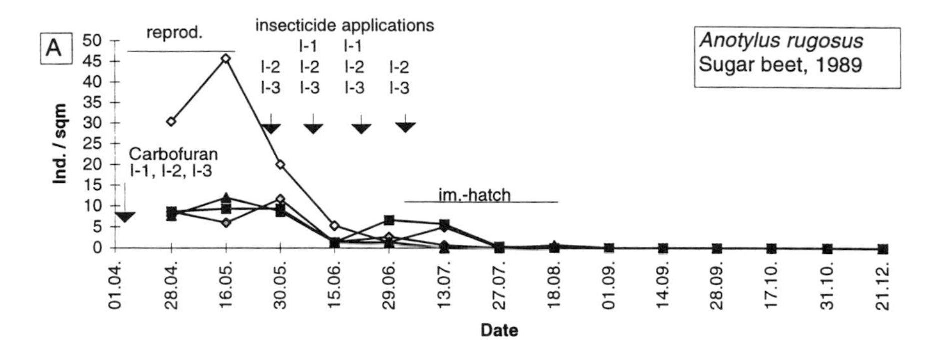

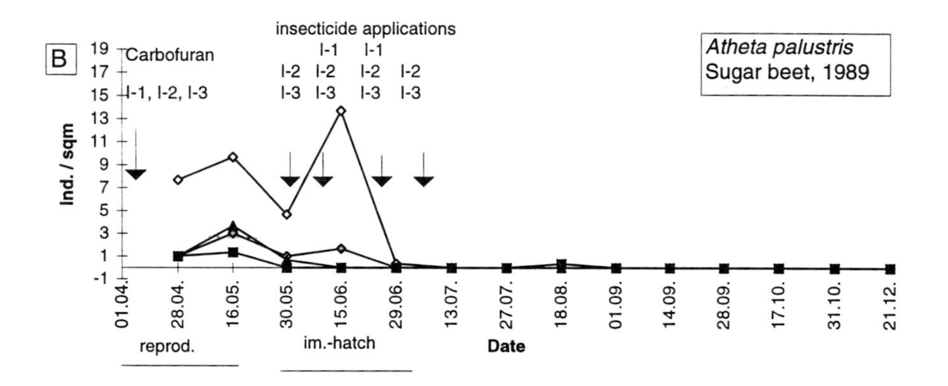

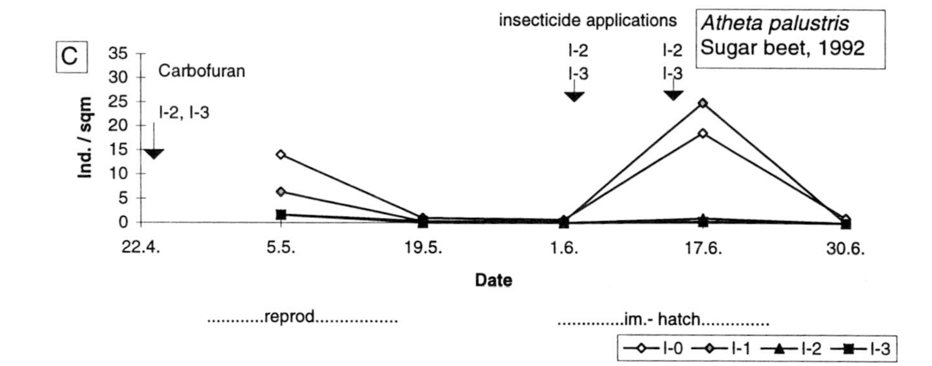

Fig. 2. Phenology of *Anotylus rugosus* (2A), *Atheta palustris* (2B and 2C), *Oxypoda exoleta* (2D), *Aloconota gregaria* (2E) and *Atheta aegra* (2F) in plots with different input of pesticides and fertilizers (I_0, I_1, I_2, I_3) with sugar beet or winter wheat in 1989, 1990 and 1992. im.- hatch = hatching of the imagines; reprod. = period of reproduction.

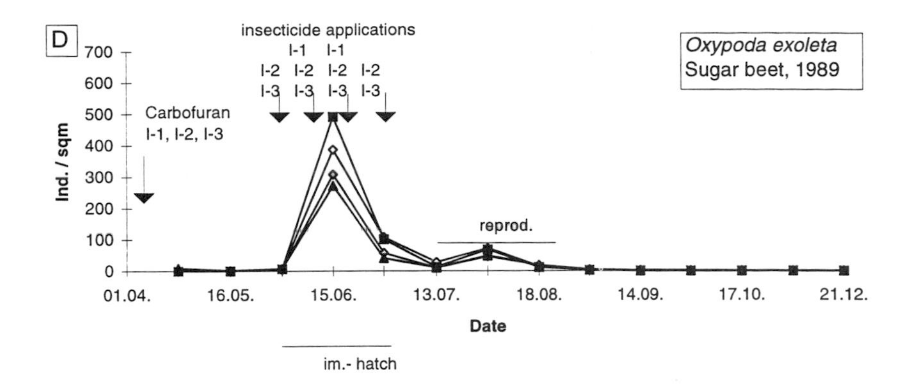

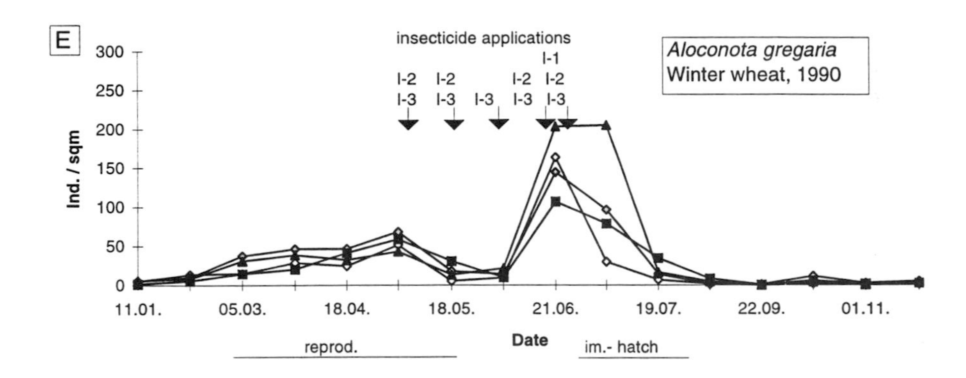

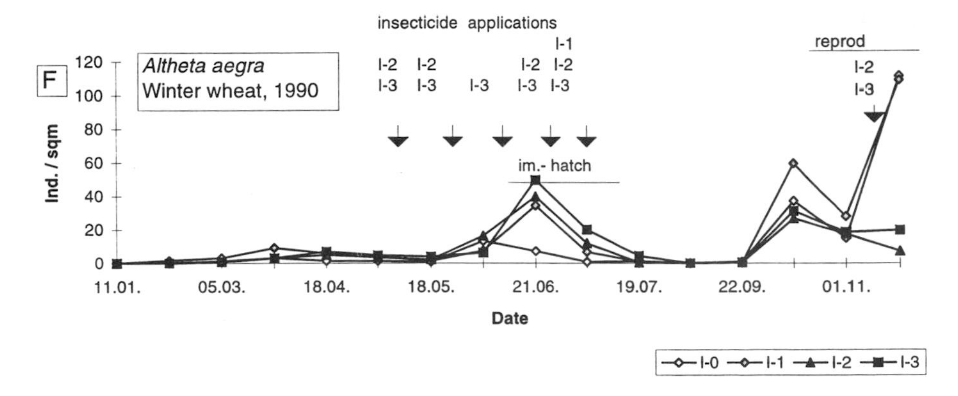

Fig. 2. (cntd) Phenology of *Anotylus rugosus* (2A), *Atheta palustris* (2B and 2C), *Oxypoda exoleta* (2D), *Aloconota gregaria* (2E) and *Atheta aegra* (2F) in plots with different input of pesticides and fertilizers (I_0, I_1, I_2, I_3) with sugar beet or winter wheat in 1989, 1990 and 1992. im.- hatch = hatching of the imagines; reprod. = period of reproduction.

I_0 than in I_1, I_2 and I_3. In addition to these findings delayed effects were recorded in the phenology of *Atheta palustris* (Fig. 2B). As described for *A. rugosus*, the treatment took place during the reproductive period and caused strong initial effects, which are recognizable by diminished abundance in the carbofuran treated plots. In addition, nearly three months after the application there was an obvious correlation between the treatments and the size of the next emerging generation. Probably the carbofuran caused lethal or sublethal effects on the reproducing population, so that less eggs were layed in the treated plots I_1, I_2 and I_3. In 1992 carbofuran-granules were applied only in the cultivation intensities I_2 and I_3 (Table 2). According to the interpretation given above, in this experiment the phenology of *Atheta palustris* (Fig. 2C) shows both, initial and delayed reduction of abundance in the treated plots. In 1989 nearly all dominant and subdominant spring breeders showed a reduced emergence in the treated plots (*Acrotona pygmaea, Amischa analis, Aloconota gregaria, Atheta elongatula, Atheta pittionii*, Fig 2). In 1992 *Atheta palustris* was the only species with a diminished emergence, which can be related to the carbofuran-application. In that year the treatment took place more than three weeks later than in 1989. The reproductive period of the spring breeders had probably finished when the insecticide was applied.

The only abundant species which lives as larva in the soil at the time of application was *Oxypoda exoleta*. This species reproduces in August and the new generation emerges in the next year in June (Fig. 2D). It is possibly this type of life-cycle which was responsible for the lack of a correlation between the treatment in March and the size of emergence of the next generation in June (Fig. 2D). The larvae, in contrast to the adult beetles, probably have no contact with the soil insecticide. It is suspected that they are less active on the soil surface or living deeper in the soil than the adults.

In 1990 winter wheat was grown. In this crop insecticides were applied against aphids during the period between May and June (Table 2). Under the influence of a gradation of aphids the cultivation intensities I_2 and I_3 were frequently treated with insecticides, partly with pesticides which are known for high initial toxicity (e.g. lambda-cyhalothrin, parathion or fenvalerate). But none of the most abundant species showed a correlation between size of emergence and insecticide treatments. For instance Fig. 2E shows the phenology of the most abundant species *Aloconota gregaria*. The first peak of emergence, which is located between February and May can be related to the period of reproduction. The second peak, between mid-June and mid-July, indicated hatching of the imagines of the new generation. The applications took place when reproduction had already finished. At that time the popu-

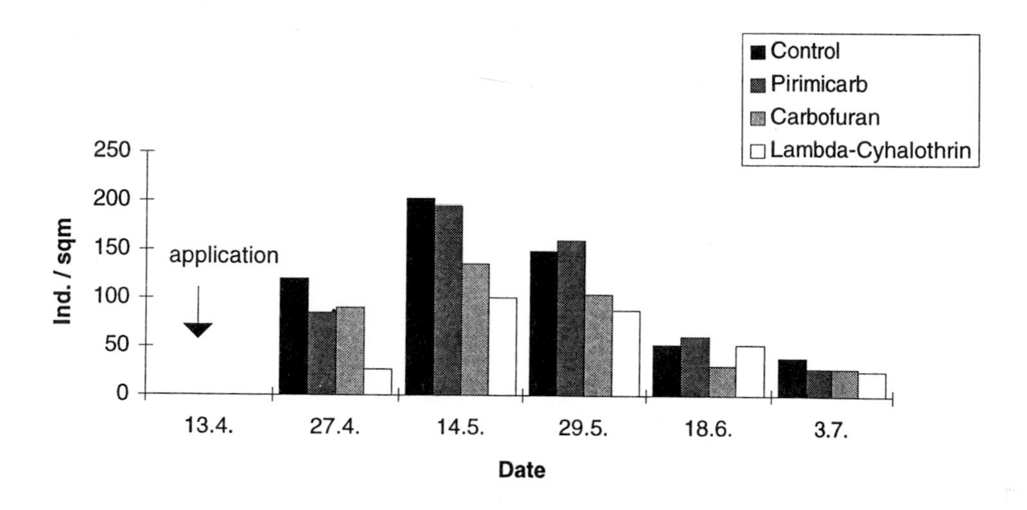

Fig. 3. Phenology of Staphylinidae in plots treated with different insecticides. Crop: sugar beet, 1993

lation lived as larvae in the soil, so that no direct contamination could occur. Even during the period of hatching of the imagines *Aloconota gregaria* was obviously not effected by the treatments. Presumably the new generation tends to leave the field quickly after hatching. Only plot I_3 shows a slightly diminished emergence. However, the highest numbers of emergence were measured in I_2, which was treated frequently with insecticides as often as I_3. The phenologies of all other species which are not reproducing during the period of insecticide treatments, might be interpreted in the same way.

After harvesting the winter wheat in August, winter barley was sown in September. In this crop an exceptional fenvalerate application against aphids was carried out in the cultivation intensities I_2 and I_3 in autumn (23rd October) because of the BYDV risk. Effects of this application were recorded only in autumn breeders. For instance Fig. 2E shows the phenology of *Atheta aegra*. This species reproduces in autumn and winter time, while the hatching of imagines occurs in June. The fenvalerate-application considerably reduced abundance in the treated plots. The phenologies of other autumn breeders can be interpreted in a similar way (e.g. *Lesteva longelytrata* and *Oxypoda longipes*).

To confirm the results described above, in 1993 an additional field trial was carried out to measure especially the effects of the soil insecticide carbofuran. For that reason four different plots each with 100 square meter were marked and treated on the 13th April. Besides carbofuran two more insecticides (pirimicarb and lambda-cyhalothrin) were tested as a comparison to carbofuran. The beetles were sampled by four emergence traps per plot, which were not moved during ten weeks (the period of the experiment). In Fig. 3 the phenology of staphylinid beetles is plotted. As expected, pirimicarb seems to be relatively harmless to the beetles. On the other hand, lambda-cyhalothrin shows the strongest initial and longer-lasting effects. Shortly after the treatments the abundances in the carbofuran treated plot were only slightly diminished compared to the untreated control. The toxicity of the soil insecticide was higher four to eight weeks after application.

Discussion

The field trial was designed for measuring the effects of different cultivation intensities on the rove-beetles. The four plots differ mainly in the input of nitrogen, pesticides and the density of weeds. All those factors strongly influenced abundance and emergence of the rove-beetles. Although the field trial was not designed to measure the effects of a certain pesticide, some correlations between treatments and abundance were discovered. Insecticide treatments, which took place during the period of larval development and during the hatching of imagines produced no measurable effects. Larvae living in the soil are protected from direct contamination by the pesticides. After hatching the specimens are not affected because they leave the field very quickly for dispersion flights. As Sprick (1991) already suggested, the scale of effects of pesticides should not be deduced from the abundance of species hatching during the application period. In this period an assessment of toxicity of pesticides leads to an over optimistic evaluation.

However, for these kind of tests the results presented here indicated that populations reproducing during the treatments and those species which are active on the soil surface are most sensitive to pesticides. In 1989 an application of carbofuran granules relatively early in the year during the reproductive period of many species reduced the abundance of most spring breeders. Similar observations are published by Heimbach & Abel (1991). In addition, for *Atheta elongatula, Atheta palustris* and *Atheta pittionii* delayed effects occurred. The treatment obviously caused a reduction of the number of eggs laid (Zimmermann & Büchs 1993, Zimmermann 1995). The effect was measurable three months later in summer by a diminished emergence rate of the second generation. Clements et al. (1988) and Kokta (1989) suspected similar effects for some species of carabid beetles.

Furthermore, exceptional autumn treatments caused effects in autumn breeders (*Atheta aegra, Lesteva longelytrata, Oxypoda longipes*), which are usually unaffected by the normal management of the field crops, even if there are many insecticide applications during the summer.

In intensively managed field crops, the rove-beetle coenosis seems to be adapted to the frequent use of insecticides. Species reproducing during the period of applications (May to June) are relatively under-represented. In contrast the abundances of species reproducing either earlier or later in the year are relatively high. To support at least the populations of beneficial rove-beetles, a crop management regime should minimize insecticide applications in early spring and autumn.

References

Assing, V. 1993. Zur Bionomie von *Xantholinus rhenanus* Coiff. und anderen bodenbewohnenden Xantholinen (Col., Staphylinidae). *Zool. Jb. Syst.* 120: 13-38.

Büchs, W. 1993. Auswirkungen unterschiedlicher Bewirtschaftungsintensitäten auf die Arthropodenfauna von Winterweizenfeldern. *Verh. Ges. Ökol.* 22: 27-34.

Büchs, W. 1994. Effects of different input of pesticides and fertilizers on the abundance of arthropods in a sugar beet crop: an example for a long-term risk assessment in the field. In: Donker, M., Eijsackers, H. & Heimbach, U. (eds.). *Ecotoxicology of soil organisms*: 303-19, Lewis Publishers, Boca Raton.

Clements, R.O., Asteraki, E. & Jackson, C.A. 1988. A method to study the effects of chlorpyriphos on predatory ground beetles in grassland. *BCPC Mono.* Nr. 40 (Field methods for the study of environmental effects of pesticides): 167-74.

Eghtedar, E. 1970. Zur Biologie und Ökologie der Staphyliniden *Philonthus fuscipennis* Mannh. und *Oxytelus rugosus* Grav. *Pedobiol.* 10: 169-79.

Hartmann, P. 1979. Biologisch-ökologische Untersuchungen an Staphylinidenpopulationen verschiedener Ökosystemen des Solling. Ph.D. Thesis, Univ. Göttingen.

Heimbach, U. & Abel, C. 1991. Nebenwirkungen von Bodeninsektiziden in verschiedenen Applikationsformen auf einige Nutzarthropoden. *Verh. Ges. Ökol.* 19: 163-70.

Kasule, F.K. 1968. Field studies on the life histories of some British Staphylinidae (Coleoptera). *Trans. Soc. Br. Entomol.* 18: 49-80.

Kasule, F.K. 1970. Field studies on the life histories of *Othius (Gyrohypnus auctt.)*

punctulatus (Goeze) and *O. myrmecophilus* (Kiesenwetter) (Coleoptera: Staphylinidae). *Proc. R. Ent. Soc. Lond.* A. 45: 55-67.

Kokta, C. 1989. Auswirkungen abgestufter Intensität der Pflanzenproduktion auf epigäische Arthropoden, insbesondere Laufkäfer (Coleoptera, Carabidae) in einer dreigliedrigen Fruchtfolge. Ph.D. Thesis, T.H. Darmstadt.

Lipkow, E. 1966. Biologisch-ökologische Untersuchungen über Tachyporus-Arten und *Tachinus rufipes* (Col., Staphylinidae). *Pedobiol.* 6: 140-77.

Schimke, G. 1978. Einfluss von Temperatur und Photoperiode auf Entwicklung und Diapause einiger Staphyliniden. *Pedobiol.* 18: 1-21.

Sprick, P. 1991. Erfassung und Beurteilung der Nebenwirkungen von Pflanzenschutzmitteln auf epigäische Coleopteren unter besonderer Berücksichtigung der Eignung von Bodenfallen. Ph.D. Thesis, Univ. Hannover.

Topp, W. 1979. Vergleichende Dormanzuntersuchungen an Staphyliniden (Coleoptera). *Zool. Jb. Syst.* 106: 1-49.

Zimmermann, J. & Büchs, W. 1993. Verzögerte Einflüsse von Pflanzenschutzmitteln auf die Kurzflügelkäfer (Coleoptera: Staphylinidae) eines Zuckerrübenfeldes unter besonderer Berücksichtigung eines Bodeninsektizides. *Verh. Ges. Ökol.* 20: 183-1990.

Zimmermann, J. 1995. Biologisch-ökologische Untersuchungen an Kurzflügelkäfern (Coleoptera: Staphylinidae) unterschiedlich intensiv bewirtschafteter Agrarflächen unter Berücksichtigung methodischer und ökotoxikologischer Gesichtspunkte. Ph.D. Thesis, TU Berlin.

Effects of reduced tillage systems in sugar beet on predatory and pest arthropods

U. Heimbach & V. Garbe

Federal Biological Research Centre for Agriculture and Forestry,
Institute for Plant Protection in Arable Crops and Grassland,
Messeweg 11/12, 38104 Braunschweig, Germany

Abstract
Two years of field trials in sugar beet on different sites comparing conventional and reduced tillage methods with *Sinapis alba* and *Phacelia tanacetifolia* as catch crops indicated that the number of predatory arthropods (spiders, rove and carabid beetles) increased drastically, especially if conservation tillage without seed-bed preparation was carried out. When *Sinapis* was used as a catch crop higher numbers of predators were found compared to *Phacelia*. In plots without seed-bed preparation more predators, especially spiders, were caught than in plots with seed-bed preparation. No clear interpretation of the reasons for differences between treatments can be given yet. Predators have to be analysed to species level and potential prey has to be monitored.

The damage caused by pest insects to the sugar beet was reduced in conservation tillage plots. About 20% less aphid-infested plants were found and biting damage to sugar beet seedlings was reduced to about 50% of that in the conventional system. The sugar yield, calculated for two experiments, was not different between the treatments. Thus, conservation tillage in sugar beet can protect the environment without reducing the yield of the crop.

Key words: carabids, spiders, rove beetles, *Sinapis alba, Phacelia tanacetifolia*, conservation tillage

Introduction

Conservation tillage is used to avoid erosion (Unger & McCalla 1980) and to reduce leaching of nitrogen. However, changing the soil tillage system can also influence the population dynamics of animals, especially soil organisms, in the field. Thus it is well known that ploughing can influence earthworm

Arthropod natural enemies in arable land · II *Survival, reproduction and enhancement*
C.J.H. Booij & L.J.M.F. den Nijs (eds.). *Acta Jutlandica* vol. 71:2 1996, pp. 195-208
 ISBN 87 7288 672 2

numbers quite drastically (Friebe & Henke 1991, Schwerdle 1969) and plant residue decomposition increases when tillage intensity is reduced (Friebe & Henke 1991). Direct drilling systems or reduced tillage systems quite often had positive effects on predatory arthropod abundance. Most of the experiments were carried out in the United States (Blumberg & Crossley 1983, Brust & House 1990, House 1989, House & Stinner 1983, Weiss et al. 1990, Stinner et al. 1986) and only a few in western Europe (Garbe & Heimbach 1992, Kendall et al. 1991, Paul 1986). Provision of additional organic material influences the numbers of saprophagous arthropods, can have effects on predatory arthropods (Humphreys & Mowat 1994, Purvis & Curry 1984) and may provide good overwintering sites for arthropods which are lacking in arable fields (Wratten & Thomas 1990). It is also known that the addition of other plant material to the soil surface, such as by undersowing, or mulching, or the use of direct drilling systems may reduce aphid numbers and virus transmission by aphids (Finch & Edmonds 1994, Jones 1994, Kendall et. al. 1991, Paul 1986), as well as other pest insect damage (Brust & House 1990, House & Stinner 1983, Stinner et al. 1986).

The aim of our experiments was to investigate the influence of direct drilling and also of conservation tillage in sugar beet, using *Phacelia tanacetifolia* and *Sinapis alba* as catch crops, on the number of predatory arthropods. The incidence of pest insects and damage to sugar beet plants were also monitored. The initial results of some of the experiments were published by Garbe & Heimbach (1992).

Materials and Methods

In 1991 and 1992 experiments were carried out in different sugar beet fields in Hötzum and Dettum (near to Braunschweig, Lower Saxony). Two fields were used in 1991 and one in 1992. The fields were divided into plots (2 plots per treatment) of 12 to 24 m width and about 100 m length. Five different types of soil cultivation were used:

Conventional system: harvest of the previous crop of winter barley, ploughing in late autumn, soil tillage just before drilling of the sugar beet in spring.

Phacelia tillage: ploughing soon after the harvest of winter barley, seed-bed preparation and drilling of *Phacelia tanacetifolia* as a catch crop, seed-bed preparation just before drilling of the sugar beet in spring.

Phacelia non-tillage: ploughing soon after the harvest of winter barley, seed-bed preparation and drilling of *Phacelia tanacetifolia* as a catch crop, direct drilling of the sugar beet in spring without soil tillage.

Sinapis tillage: ploughing soon after the harvest of winter barley, seed-bed preparation and drilling of *Sinapis alba* as a catch crop, seed-bed preparation just before drilling of the sugar beet in spring.
Sinapis non-tillage: ploughing soon after the harvest of winter barley, seed-bed preparation and drilling of *Sinapis alba* as a catch crop, direct drilling of the sugar beet in spring without soil tillage.

A small experiment, with plots of only 12 x 25 m (without replicates), comparing only conventional tillage to *Sinapis* non-tillage was carried out in 1992 in Salzgitter (near to Braunschweig). In 1992 in Hötzum, *Phacelia* developed less well than in 1991 because of a dry period during the germination of the seeds. Therefore seeds from the previous crop, winter barley, germinated and reached about 50% soil coverage.

In all years the catch crops were killed by frost periods during winter. All herbicide and insecticide treatments were carried out in the same way on all plots, though more weeds developed in the conventional and tillage systems (Garbe & Heimbach 1992). No insecticides were used, except for pirimicarb, which is known to be harmless to most predacious arthropods. In areas of the plots where the density of arthropods and damage to sugar beet plants were monitored, no insecticidal seed dressing was used. This was done to avoid toxic influences of insecticides on the fauna.

The density of predatory arthropods was measured using ground-photo-eclectors (emergence traps), of area 0.25 m², with a pitfall trap buried inside the eclector. The eclectors were emptied and moved to a new location once or twice a week, thus providing population densities over a range of dates. The eclectors were placed on the field immediately after the drilling of sugar beet (and in one experiment even before that) until about the end of June. At that time the soil cover by sugar beet plants was about 100%. The eclectors were always placed in the same row of beet plants, covering usually two plants per eclector. Usually two eclectors per plot (Dettum, 1991 only one) were used.

The use of photoeclectors gives quite valuable results and has an advantage over pitfall trapping that the results are not so biased by the activity of arthropods. Eclectors are quite efficient for catching spiders and rove beetles (cf. D-Vac, spiders only). A recent and valuable discussion of the methods is given by Sunderland et al. (1995) and Volkmar et al. (1994). Carabids were only caught in low numbers because of the small sampling area and because there were few in sugar beet fields. In addition to the fact that pitfalls are greatly influenced by activity, it was also not appropriate to use them because the plot size was so small.

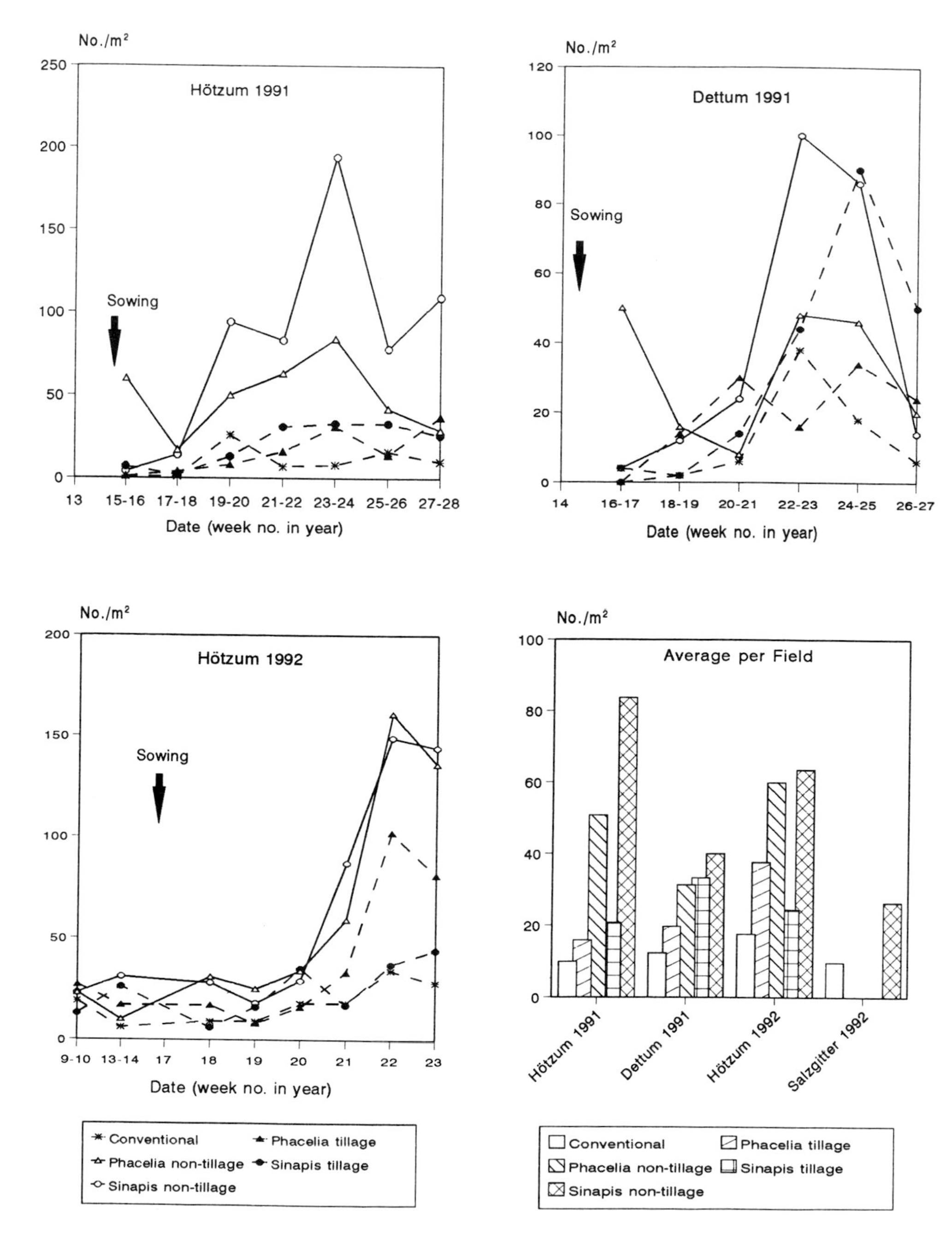

Fig. 1. Number of spiders sampled by ground-photoeclectors in conventional sugar beet fields and in plots of different catch crop with (tillage) or without (non-tillage) seed-bed preparation when drilling sugar beet.

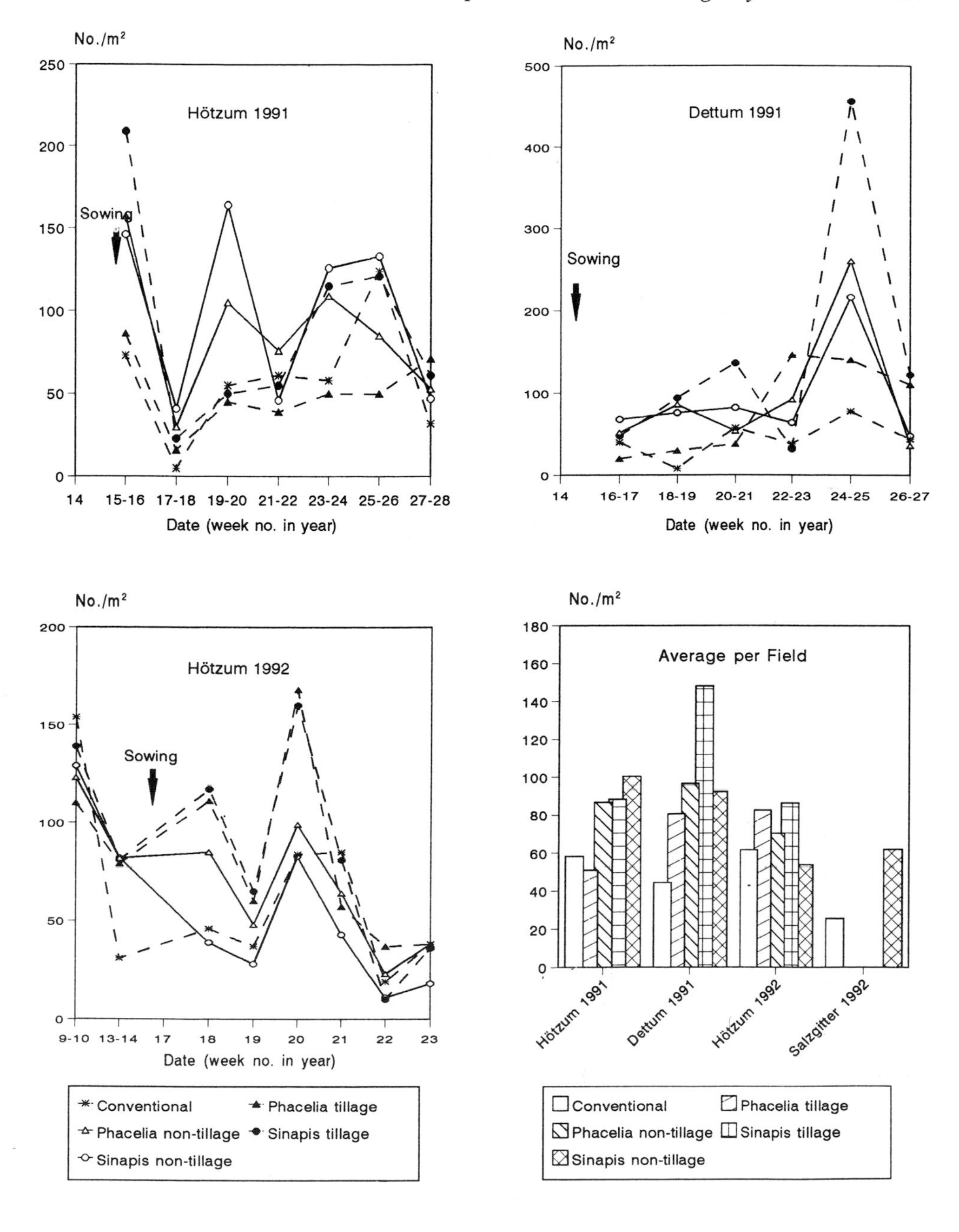

Fig. 2. Number of staphylinid beetles sampled by ground-photoeclectors in conventional sugar beet fields and in plots of different catch crop with (tillage) or without (non-tillage) seed-bed preparation when drilling sugar beet.

To check the damage to the beet seedlings ten seedlings per plot were analysed (ES 13-21) and the number of wounds caused by pest insects such as *Onychiurus* spp., and *Atomaria linearis* were counted. At different dates the number of aphid-infested plants (25-50 plants per plot) were counted (ES 21-25). Aphid counting was carried out before treatments with pirimicarb were necessary.

In 1991 sugar beets were harvested in small subplots in all plots and the yield and sugar content were analysed. Results on the time and percentage of germination of the sugar beet seedlings, as well as on weed development in the different plots, are given by Garbe & Heimbach (1992).

Results

About 90% of all beneficial arthropods collected in the eclectors were epigeic predatory arthropods belonging to the groups Araneae, Staphylinidae and Carabidae.

The number of spiders (Fig.1) increased slightly during the season; they differed little between the different tillage systems in the first weeks after drilling of the beets but thereafter reached very high numbers, especially in the system with no seed-bed preparation in spring. They peaked at more than 100 spiders per m². The average values show very clearly the distinct differences between the tillage systems. *Sinapis* as a catchcrop seemed to have a more positive influence on spiders than *Phacelia*, and distinctly higher numbers occurred in non-tillage systems than in tillage systems.

The number of rove-beetles (Fig. 2) was quite high even just before and after drilling of the beets in April. The numbers maintained a similar level during the whole period in Hötzum in 1991 (50-150/m²), seemed to decrease slightly during the season in Hötzum in 1992 (from 125 to 30/m²) and to increase in Dettum in 1991. On average staphylinid beetle density was higher in the reduced tillage system plots in nearly all cases. No clear effects were found on beetle numbers in plots with seed-bed preparation in spring. In 1991 in both fields higher numbers of beetles were found in *Sinapis* plots, but in 1992 this happened in *Phacelia* plots.

The number of carabids (Fig. 3) during the season is given only for the field with the highest carabid density. The numbers increased during the growing period from less than 10 to more than 30/m² in plots with catch crops, but never exceeded 16/m² in the conventional plot. Clearly more carabids were caught when *Sinapis* was used as a catch crop compared to *Phacelia,* and also in plots with no seed-bed preparation (for both types of

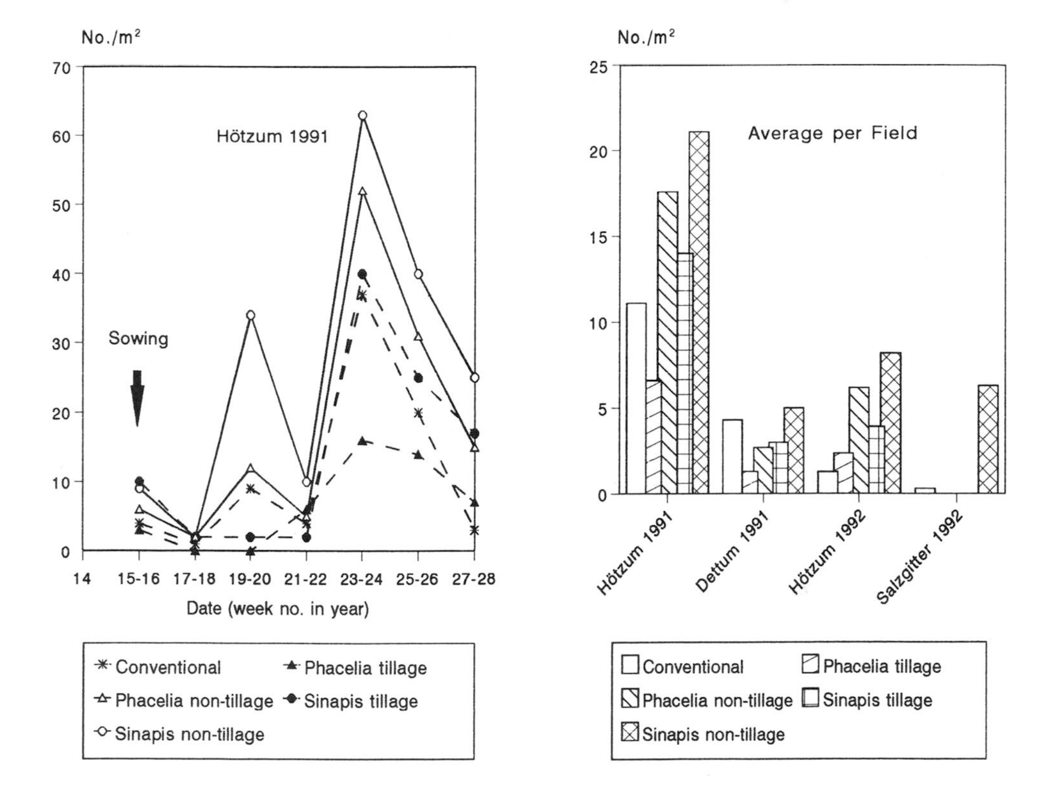

Fig. 3. Number of carabid beetles sampled by ground-photoeclectors in conventional sugar beet fields and in plots of different catch crop with (tillage) or without (non-tillage) seed-bed preparation when drilling sugar beet.

catch crop.) Carabids in the conventional system reached densities which were usually lower than in the plots with catch crops. A similar density to that in *Sinapis* non-tillage plots was found in only one field.

The number of wounds per sugar beet seedling (Fig. 4) was distinctly reduced in nearly all plots and all years when a catch crop had been used. There was a tendency for less bites per plant when *Sinapis* was used as the catch crop, compared to *Phacelia,* but no clear difference was found between tillage and non-tillage systems. The percentage of aphid-infested (*Myzus persicae* and *Aphis fabae*) plants (Fig. 5) was lower in all catch crop plots of the three fields compared to the conventional system (except in one case). Numbers of aphid-infested plants were especially reduced in Hötzum in 1992 in plots with *Phacelia* whereas in both fields in 1991 *Sinapis* plots had lower numbers. Seed-bed preparation in spring did not seem to have a large influence on aphid infestation, but there was a tendency for lower numbers in plots without seed-bed preparation. Altogether the number of sugar beets per m^2 and the development of the crop differed only to a small degree between the systems (Garbe and Heimbach 1992) and the sugar yield (Fig. 6) was very similar in all plots.

Discussion

The use of catch crops versus the conventional sugar beet growing system may influence arthropods in different ways.

Firstly, if a catch crop is used, the time of ploughing of the field after harvest of the previous crop (in our experiments winter barley) is about two months earlier than in the conventional system. This may affect species which are sensitive to mechanical influences at the time of ploughing. Our own laboratory experiences indicate that larvae of carabids and staphylinids are sensitive to the type of soil structure and to disturbance of the soil. Adults do not seem to be killed directly by soil tillage (Lorenz et al. 1994).

Secondly, catch crops keep the field green over the winter period and may therefore supply predatory arthropods with shelter and overwintering sites as well as with an improved food supply resulting from increases in numbers of saprophagous and phytophagous insects. In general, the activity of organisms on the soil surface is increased when the field is covered with plants during winter compared to bare soil in the conventional system. These effects may vary due to the type of catch crop and the extent to which it augments food supplies. For example, in our experiments, aphids and their specific antagonists were found in autumn only in *Sinapis* but not in *Phacelia* plots.

Thirdly, more decaying organic material is available on the soil surface after the drilling of sugar beet in spring. This provides food for saprophagous arthropods that may be a food resource for predators and also results in different soil and soil surface structures. The soil surface is more structured in direct drilled plots than in plots with seed-bed preparation. The soil structure provides hiding and living places for arthropods and also influences the soil temperature, which is higher in spring in the conventional system. Changes of the micro-habitat can have large effects on the abundance of predators (Alderweireldt & Desender 1990).

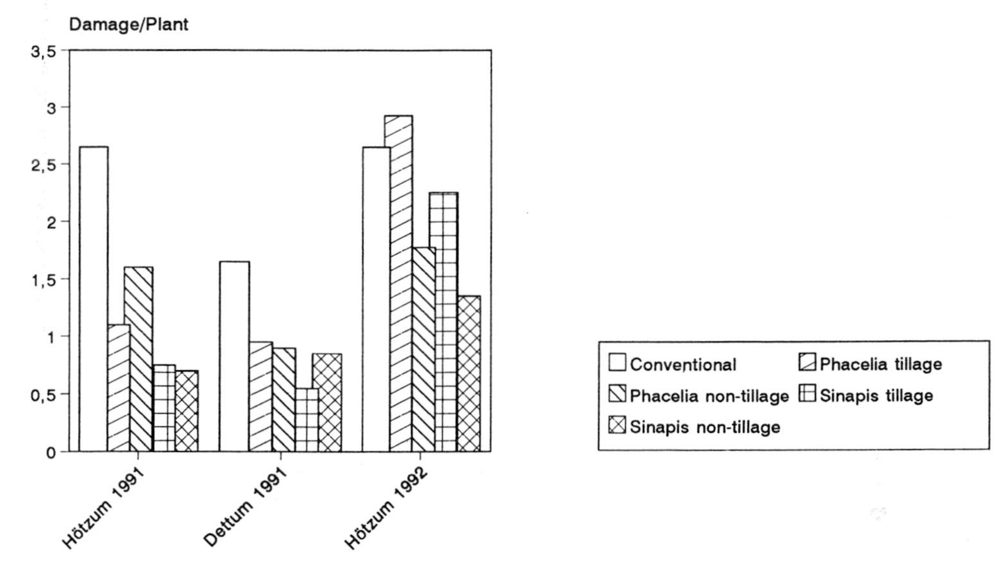

Fig. 4. Number of biting points by Collembola and *Atomaria* on sugar beet seedlings in conventional fields and in plots of different catch crop with (tillage) or without (non-tillage) seed-bed preparation when drilling sugar beet.

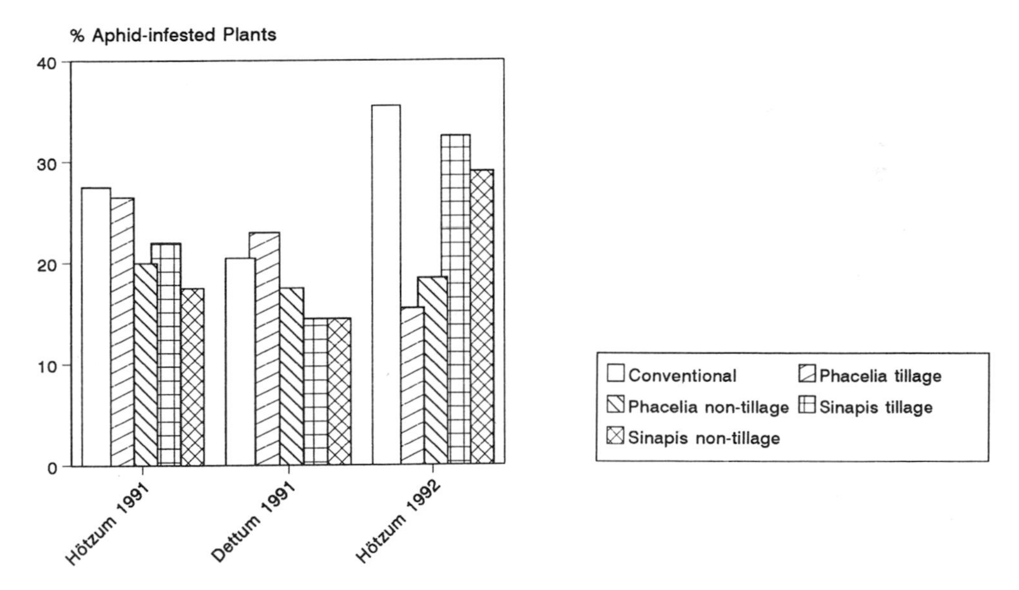

Fig. 5. Percentage of aphid-infested sugar beet plants in conventional fields and in plots of different catch crop with (tillage) or without (non-tillage) seed-bed preparation when drilling sugar beet.

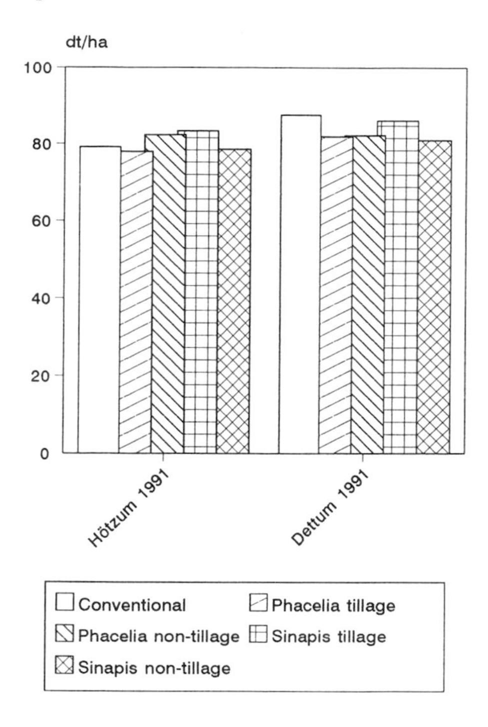

Fig. 6. Yield of sugar in conventional sugar beet fields and in plots of different catch crop with (tillage) or without (non-tillage) seed-bed preparation when drilling sugar beet.adult/juvenile stages.

Our results on numbers of predatory arthropods need to be analysed more deeply by separating species, sexes and life stages.

Additionally the number of important potential food items for the predators (e.g. Collembola and Diptera) have to be monitored to get to a better understanding of population density differences between the treatments. Further research is needed to come to clear conclusions.

Spiders were very sensitive to the different tillage systems. This may be due to high numbers of potential prey items that are supported by the organic material available but also to a more structured habitat giving more or better web building sites which can influence spider density (Rypstra 1983). Because *Sinapis* left more organic material and structures on the soil surface than *Phacelia*, this may be the reason for more spiders in *Sinapis* plots. But more detailed analysis of the material is necessary to find out which species increased in number and whether juveniles or adults were responsible for the increase of spiders in plots without seed-bed preparation later in the season. Seed-bed preparation before the drilling of sugar beet may affect spiders mechanically, but at that time of the season numbers are too low to analyse this effect. No clear differences in spider numbers occur between the treatments early in the season. Though numbers are quite low at that time it does not seem that overwintering sites in the catch crop are of importance. But it is well known that spiders move easily between habitats by ballooning (Sunderland 1991).

Numbers of staphylinid beetles also increased when reduced tillage systems were used. It is much less clear than in the case with spiders in which way seed-bed preparation affects the density of these beetles and why it did not affect them in Hötzum in 1992. Even if only the data from the period after the drilling of the sugar beet are analysed, the *Sinapis* non-tillage system gives the lowest values. The different quality of overwintering sites did not result in different beetle numbers before the drilling of sugar beets in Hötzum in 1992. Seed-bed preparation may affect larval stages and result in lower numbers later in the year.

Though the number of carabids is too low and therefore also too variable to permit rigorous conclusions, similarities to spider numbers can be noted. *Sinapis* is better than *Phacelia* and seed-bed preparation worse than direct drilling into the catch crop. For carabids a larger area has to be monitored, especially if interpretation is to be done at the species level. As for staphylinids seed-bed preparation may affect larval stages and result in lower carabid numbers later in the year. In the experiments by Hance & Gregoire-Wibo (1987) the season in which the plots were ploughed affected carabid beetles differently, and bare soil during winter had negative effects on beetle numbers and species diversity.

Numbers of saprophagous insects increase if organic material is added (House, 1989). *Onychiurus* spp. and *Atomaria linearis* can damage sugar beet seedlings but Collembola also feed on decaying organic material (Ulber, 1980). Thus reduced damage to the crop, in plots with catch crops, as was found in our experiments, may be due to a higher number of predators or to the availability of food items that provide an alternative to sugar beet seedlings. This may also be the reason for slightly lower damage in *Sinapis* plots compared to *Phacelia* as there are more organic material and more predators in *Sinapis*. To differentiate between these effects the numbers of pest insects will have to be monitored. Other experiments have shown that less damage to sugar beet seedlings results if alternative organic material is available for the pests (Garbe 1987, Sievers & Ulber 1990).

In our experiments we had less aphid-infested plants in reduced tillage plots. This may be due to the increase in predatory arthropod numbers and also due to a changed settling behaviour of aphids. Aphids react differently to bare soil with a few contrasting green areas of sugar beet plants compared with a soil surface covered with decaying organic material which has less colour contrast. Jones (1994) also found less aphids and virus transmission in lupine in plots with mulching of straw and concluded that the settling of aphids was influenced by this. Other authors found less aphids in undersowing systems (Finch & Edmonds 1994), mulching in cereals (Kendall et al. 1991) or when some weed plants were left in sugar beet (El Titi 1986). Thus aphid behaviour and predator-prey interactions have to be studied in more detail before final conclusions can be drawn.

Acknowledgement

The authors would like to thank Ms. C. Langenstück and I. Rosenbruch for their help during the field experiments, farmer E. Curland for supplying one of his fields and Ms. A. Wehling for her help in preparing the manuscript and Dr. K.D. Sunderland for critical comments and correcting the English.

References

Alderweireldt, M. & Desender, K. 1990. Microhabitat preference of spiders (Araneae) and carabid beetles (Coleoptera, Carabidae) in maize fields. *Med. Fac. Landbouww. Rijksuniv. Gent* 55: 501-10.

Blumberg, A.Y. & Crossley, D.A. 1983. Comparison of soil surface arthropod populations in conventional tillage, no-tillage and old field systems. *Agro-Ecosystems* 8: 247-53.

Brust, G.E. & House, G.J. 1990. Effects of soil moisture, no-tillage and predators on southern corn rootworm (*Diabrotica undecimpunctata howardi*) survival in corn agroecosystems. *Agric. Ecosyst. & Environ.* 31, 199-215.

El Titi, A. 1986. Zum ökonomischen Nutzen von Ackerunkräutern im integrierten Pflanzenschutz, dargestellt am Zuckerrübenanbau. Proc. EWRS Symp. 1986, *Economic Weed Control*, 209-16.

Finch, S. & Edmonds, G.H. 1994. Undersowing cabbage crops with clover – the effects on pest insects, ground beetles and crop yield. *IOBC/WPRS Bull.* 17(8): 159-67.

Friebe, B. & Henke, W. 1991. Bodentiere und deren Strohabbauleitung bei reduzierter Bodenbearbeitung. *Z. Kulturtech. und Landentwickl.* 32: 121-26.

Garbe, V. 1987. Verunkrautung und Auftreten von Schädlingen bei unterschiedlichen Systemen der Bodenbearbeitung zu Zuckerrüben. Dissertation, University Göttingen, 108 pp.

Garbe, V. & Heimbach, U. 1992. Mulchsaat zu Zuckerrüben. *Zuckerrübe* 41: 230-34.

Hance, T. & Gregoire-Wibo, C. 1987. Effect of agricultural practices on carabid populations. *Acta Phytopathol. Entomol. Hungarica* 22: 147-60.

House, G.J. & Stinner, B.R. 1983. Arthropods in no-tillage soybean agroecosystems: community composition and ecosystem interactions. *Environ. Management* 7: 23-28.

House, G.J. 1989. No-tillage and legume cover cropping in corn agroecosystems: Effects on soil arthropods. *Acta Phytopathol. Entomol. Hungarica* 24: 99-104.

Humphreys, I.C. & Mowat, D.J. 1994. Effects of organic treatment on beetle predators of the cabbage root fly and on alternative prey species. *IOBC/WPRS Bull.* 17(8): 115-23.

Jones, R.A.C. 1994. Effect of mulching with cereal straw and row spacing on spread of bean yellow mosaic potyvirus into narrowleafed lupins (*Lupinus angustifolius*). *Ann. appl. Biol.* 124: 45-58.

Kendall, DA., Chinn, N.E., Smith, B.D., Tidboald, C., Winstone, L. & Western, N.M. 1991. Effects of straw disposal and tillage on spread of barley yellow dwarf virus in winter barley. *Ann. appl. Biol.* 119: 359-64.

Lorenz, E., Ulber, B. & Poehling, H.M. 1994. Einfluß verschiedener mechanischer Unkrautbekämpfungsverfahren auf die Abundanz von Laufkäfern (Coleoptera, Carabidae) in Zuckerrüben. *Z. Pflanzenkrankh. Pflanzensch. Spec.* 14: 635-44.

Paul, W.D. 1986. Vergleich der epigäischen Bodenfauna bei wendender bzw. nichtwendender Grundbodenbearbeitung. *Mitt. Biol. Bundesanst. Land-Forstwirtsch.*, Berlin Dahlem 232: 290 pp.

Purvis, G. & Curry, J.P. 1984. The influence of weeds and farmyard manure on the activity of Carabidae and other ground-dwelling arthropods in a sugar beet crop. *J. appl. Ecol.* 21: 271-83.

Rypstra, A.L. 1983. The importance of food and space in limiting webspider densities; a test using field enclosures. *Oecologia* 59: 312-16.

Schwerdle, F. 1969. Untersuchungen zur Populationsdichte von Regenwürmern bei herkömmlicher Bodenbearbeitung und bei "Direktsaat". *Z. Pflanzenkrankh. Pflanzensch.* 76: 635-41.

Sievers, H. & Ulber, B. 1990. Freilanduntersuchungen zu den Auswirkungen der organischen Düngung auf Collembolen und andere Kleinarthropoden als Auflaufschädlinge in Zuckerrübenbeständen, *Z. Pflanzenkrankh. Pflanzensch.* 97: 588-99.

Stinner, B.R., Krueger, H.R. & McCartney, D.A. 1986. Insecticde and tillage effects on pest and non-pest arthropods in corn agroecosystems. *Agricult. Ecosyst. & Environ.* 15: 11-21.

Sunderland, K.D. 1991. The ecology of spiders in cereals. *Proc.6th Internat.Symp. Pests & Diseases*, Halle/Saale, Germany 269-80.

Sunderland, K.D., De Snoo, G.R., Dinter, A., Hance, T., Helenius, J., Jepson, P., Kromp, B., Lys, I.A., Samu, F., Sotherton, N.W., Toft, S., & Ulber, B. 1995. Density estimation for invertebrate predators in agroecosystems, *Acta Jutl*, in press.

Ulber, B., 1980. Untersuchungen zur Nahrungswahl von *Onychiurus fimatus* Gisin (Onychiuridae, Collembola), einem Auflaufschädling der Zuckerrübe. *Z.angew. Entomol.* 90: 333-46.

Unger, P.W. & McCalla, T.M. 1980. Conservation tillage systems. *Adv. Agron.* 33: 1-58.

Volkmar C., Bothe, S., Kreuter, T., Lübke-Al Hussein, M., Richter, L., Heimbach, U. & Wetzel, T. 1994. Epigäische Raubarthropoden in Winterweizenbeständen Mitteldeutschlands und ihre Beziehung zu Blattläusen. *Mitt. Biol. Bundesanst. Land-Forstwirtsch.*, Berlin Dahlem No.299, 134 pp.

Weiss, M.J., Balsbaugh, E.U., French, E.W. & Hoag, B.K. 1990. Influence of tillage management and cropping system on ground beetle (Coleoptera: Carabidae) fauna in the Northern Great Plains. *Environ. Entomol.* 19: 1388-91.

Wratten, S.D. & Thomas, M.B., 1990. Environmental manipulation for the encouragement of natural enemies of pests. *BCPC Mono. No.45, Organic and low input Farming*, 87-92.

Enhancement of non-target insects: indications about dimensions of unsprayed crop edges

G.R. de Snoo

Centre of Environmental Science, Leiden University, P.O. Box 9518,
2300 RA Leiden, The Netherlands

Abstract

In the Haarlemmermeer field margin project, 3 m x 100 m and 6 m x 450 m strips along the edge of a cereal crop were left unsprayed with herbicides and insecticides. By comparing these two types of unsprayed edges, an indication can be obtained of the dimensions of unsprayed crop edges required for enhancing the natural values of arable farmland. This study considers the non-target insects on the vegetation, including butterflies, hoverflies and ladybirds. Because of its intermediate role, the vegetation was also investigated. The results show that the abundance and presence of farmland plants is greatest in the outer 3 metres of the field and that there is no significant difference between a 3-metre wide edge and the outer 3 metres of a 6-metre wide edge. There is no extra increase in the density of insects and a number of selected insect groups in the 6-metre wide unsprayed edges compared with the 3-metre wide edges. Although the experimental set-up was limited in scope, it therefore appears that for these insect groups it is sufficient to leave a relatively narrow strip unsprayed.

Key words: insecticides, herbicides, butterflies, hoverflies, ladybirds

Introduction

Spray drift of pesticides to field surroundings can be markedly reduced by leaving the outer metres of the crop unsprayed (Cuthbertson & Jepson 1988). Obviously, the width of the unsprayed strip is important. At low wind speeds (3 m/s) a buffer zone of 3 metres wide is found to be sufficient to achieve a 98% reduction in the loading of the adjacent ditch (De Snoo 1994a). At these wind speeds, an unsprayed strip of 6 m width reduces pesticide deposition in the ditch by about 100%.

Arthropod natural enemies in arable land · II *Survival, reproduction and enhancement*
C.J.H. Booij & L.J.M.F. den Nijs (eds.). *Acta Jutlandica* vol. 71:2 1996, pp. 209-219
 ISBN 87 7288 672 2

By abandoning pesticide treatment on crop edges, biodiversity in farmland can also be promoted (Boatman 1994). Notable results include the return of rare species of farmland plants and the recovery of partridge populations (Schumacher 1984, Rands 1985). In addition to the question of where such unsprayed crop edges can best be realized, in terms of (landscape) ecology, it is also obviously important to find out how different dimensions contribute to benefits for nature. The width of the unsprayed strip is of great importance here, because of the pesticide input to this strip from the adjacent sprayed crop.

In the Dutch field margin project in the Haarlemmermeerpolder (1990-94) it has been investigated to which extent it is possible to enhance the natural values of arable farmland by creating unsprayed crop edges. For this purpose, both short, 3-m wide and long, 6-m wide unsprayed cereal crop edges were established and compared with the sprayed reference situation. By comparing the two types of unsprayed strips, an indication can be obtained of what dimensions unsprayed crop edges should have from the perspective of promoting biodiversity. This article discusses the impact of unsprayed crop edges on the presence and abundance of non-target insect groups (no pest-insects). Because of its intermediate role, the vegetation was also investigated. The article does not primarily consider the benefits to nature of unsprayed relative to sprayed crop edges; for a review of this topic, the reader is referred to Boatman (1994) and De Snoo & Udo de Haes (1994).

Methods

The study was carried out in the Haarlemmermeerpolder in 1992. In this polder, reclaimed about 150 years ago, most parcels are 1000 m long and 200 m wide and bordered by ditches. The clay soil contains about 23% silt (0-16 µm diam.) and 3% organic matter. The most common rotation on the farms is winter wheat followed by potatoes and a second winter wheat crop and finally sugar beet. To investigate the consequences for non-target insects, 3- and 6-m wide strips were left unsprayed with herbicides and insecticides in winter wheat. Widths of 3 and 6 m were chosen because this frequently corresponds to the length of the field sprayer boom, which can be switched off independently, permitting ready implementation of the measures in farming operations. The unsprayed 3-m wide strips were 100 m long and have remained unsprayed since 1-1-1990 (see Fig. 1). The 6-m wide strips were 450 m long and have been left unsprayed since 1-1-1992. The 6-m wide strips are mostly situated on different fields from the 3-m strips. All crop edges are parallel to a ditch.

It was investigated whether the presence and abundance of farmland plants were comparable in the various fields. First a comparison was made between farmland plants presence in the inner and outer 3 m of the 6-m unsprayed strip (ten fields). Next the presence of farmland plants in the outer 3 m of a 6-m strip was compared with that in a 3-m crop edge (established on nine fields). Finally, at the end of June, vegetation inventories were made in both the unsprayed and the sprayed edges (in the same field) at a distance of 1-2 and 4-5 m from the field edge (Braun-Blanquet method, inventory area 75 m^2 in all cases).

Fig. 1. Comparison of crop edges 3 m wide and 100 m long and crop edges 6 m wide and 450 m long.

1 = unsprayed crop edge 100 m long, 3 m wide
2 = unsprayed crop edge 450 m long, 6 m wide, outer 3 m
3 = unsprayed crop edge 450 m long, 6 m wide, inner 3 m

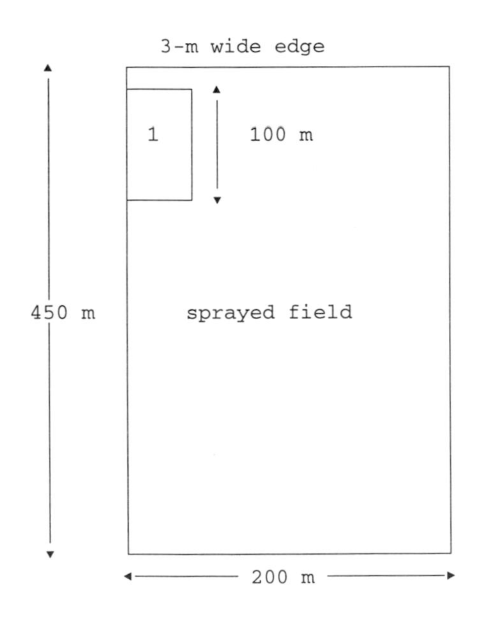

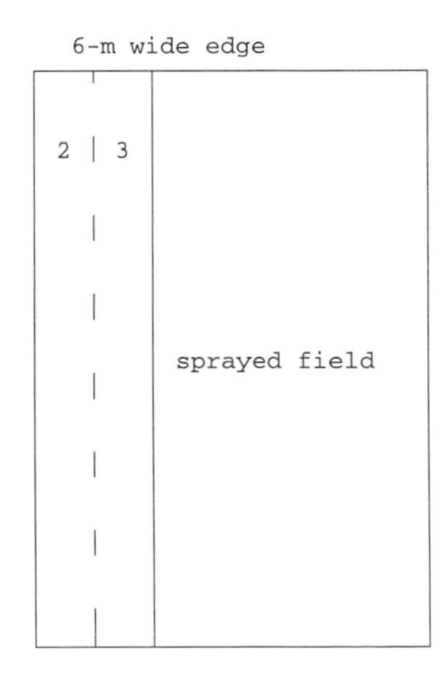

Impact on non-target insects

A butterfly census was carried out according to the transect method of Pollard (1977). Censuses were performed between 11.00 and 16.30 h at a wind speed of <10 m/s and a temperature of >17°C, or a temperature of 13-17°C and less than 40% cloud cover. Whenever possible, butterflies were inventoried once a week from 8th May to 22nd July. A total of eleven census rounds was made. The censuses were carried out on eight farms with a 3-m unsprayed crop edge and on six farms with a 6-m unsprayed edge. In each field, butterflies were also inventoried in a sprayed edge of similar width. The farms were visited in random sequence. The census areas of the crop edges were the same width as the unsprayed edges, i.e. 3 m for the 3-m edges and 6 m for the 6-m edges. To compare the number of butterflies in the 3- and 6-m unsprayed crop edges, the number of individuals in the 6-m edges was divided by two, after which the differences between both edges were tested with the Mann-Whitney U-test.

Along the crop edges, other insect groups in the upper part of the vegetation (crop and farmland plants) were sampled at the end of June. Observation took place between 10.00 and 16.00 h, at a temperature above 20°C. All the edges were sampled twice. 60 sweeps were taken per 100 m edge, with a sweep net with a diameter of 35 cm, during the same period and by the same person. The total area sampled was approx. 20 m² per 100 m. Sampling took place 1.5 m from the field edge. For mobile insect groups such as Syrphidae (hover-flies), Stratiomyidae (soldier flies), Tipulidae (crane flies), Apidae (bees and bumble-bees), Crambinae (grass moths) and Coccinellidae (ladybirds), additional visual observations were made. A total of 21 insect groups were distinguished. The Parasitica and several Diptera families were not included, because of the very small size of many representatives of these groups. The observed insects were identified at the family level or higher. A comparison was made between the presence of these insects on seven fields with a 3-m unsprayed strip and their presence in the outer 3 m of 6-m unsprayed strips (ten fields) (Mann-Whitney U-test). For the sprayed situation, too, these fields were similarly compared.

Results

Vegetation

In the outer 3 m of the 6-m wide unsprayed crop edges, the average cover of farmland plants is 35.2%, significantly higher than in the inner 3 m of the strip (25.9%, Table 1). This is also true in the sprayed situation, where these figures are 5.6 and 3.0%, respectively. A comparison between fields with unsprayed edges 3- and 6-m wide shows that the vegetation cover of the outer 3 m of the 6-m strips is very similar to that of the 3-m strips (36.2). Comparison of the presence of a number of broad-leaved species of arable land in the various parts of the crop edges gives a similar picture: in both the sprayed and unsprayed situation, the average number of such species is greater in the outer 3 m than in the inner 3 m of the 6-m wide edge, and also greater (but not significant) than in the 3-m unsprayed strip (Table 1). There is no difference between the 3-m wide edges and the outer 3 m of the 6-m wide edges. In the unsprayed crop edges, species such as *Matricaria recutita, Polgonum aviculare* and *P. convulvulus* predominate. For Graminae (e.g. *Poa annua*), finally, no significant difference was found in the presence of selected species in the outer and inner 3 m of 6-m unsprayed crop edges. Similarly, there was no difference between the 3-m crop edge and the outer 3 m of the 6-m edge (Table 1).

Table 1. Comparison of average cover of farmland plants, number of broad-leaved species and number of Graminae species (incl. standard deviation) in 3- and 6-m wide sprayed and unsprayed winter wheat edges (1992 and 1993). For positions in the field, see Fig. 1. Comparison 1-2: Mann-Whitney U-test; 2-3 Wilcoxon matched-pairs test, two-tailed. * = significant effect: $P < 0.05$; ** = significant effect: $P < 0.01$; ns - not significant: $P > 0.27$.

	1	2	3		
	3-m wide	outer 3m of	inner 3m of	test 1-2	test 2-3
	crop edge	6-m wide edge	6-m wide edge		
vegetation cover	3.9 ± 6.3	5.6 ± 12.8	3.0 ± 9.5	ns	*
sprayed edges	36.2 ± 17.0	35.2 ± 24.2	25.9 ± 23.6	ns	**
unsprayed edges					
broad-leaved species	3.4 ± 3.0	2.6 ± 2.5	1.8 ± 2.2	ns	*
sprayed edges	12.2 ± 4.3	13.4 ± 5.0	11.1 ± 6.0	ns	*
unsprayed edges					
Graminae species					
sprayed edges	2.0 ± 1.6	2.3 ± 1.0	2.1 ± 0.9	ns	ns
unsprayed edges	2.7 ± 1.1	2.6 ± 0.8	2.4 ± 0.9	ns	ns

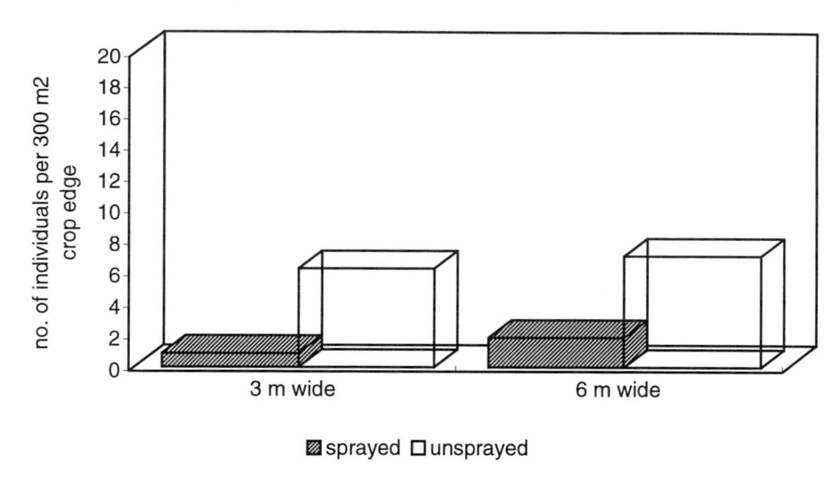

Fig. 2a. Average number of butterflies per 300 m² in 3 m and 6 m wide sprayed and unsprayed winter wheat crop edges in 1992.

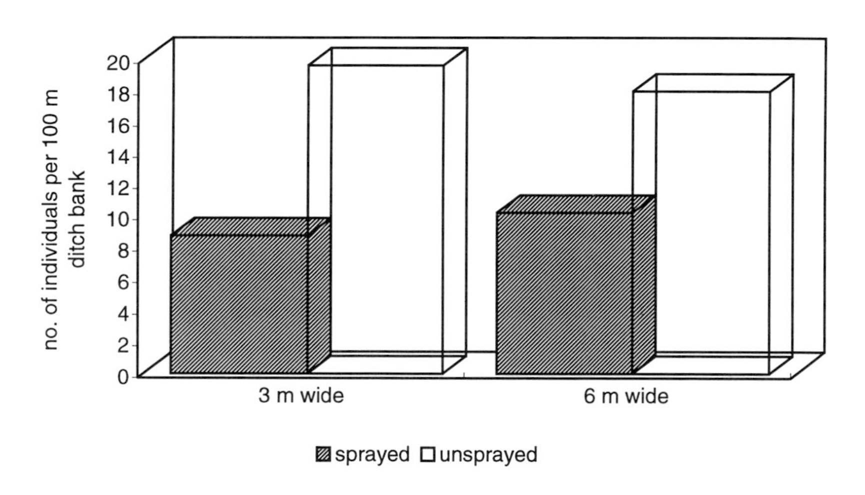

Fig. 2b. Average number of butterflies per 100 m ditch bank adjacent to 3 m and 6 m wide sprayed and unsprayed winter wheat crop edges in 1992.

Non-target insects

Comparing the unsprayed with the sprayed crop edges, the number of butterflies, the total number of other non-target insects as well as the number of insect groups increased significantly in both the 3 m wide unsprayed edges and the 6 m wide unsprayed edges, compared with the sprayed ones. However, there was great variation in insect abundance and insect species composition among the fields. Among the butterflies, Satyridae (Browns) were the most dominant family, followed by Pieridae (Whites) and Hesperiidae (Skippers). In these families, six species were common: *Maniola jurtina* (Meadow brown), *Lasiommata megera* (Wall brown), *Coenonymphha pamphilus* (Small heath), *Pieris rapae* (Small white), *P. napi* (Green-veined white) and *Thymelicus lineola* (Essex skipper). In the 6 m edges the average number of butterfly individuals (per 100 m length) is found to be about twice as high as in the 3-m edges. After conversion to abundance per unit area, however, there proves to be no significant (P >0.05) difference between the 3- and 6-m edges, whether sprayed or unsprayed. Neither is the relative increase in the number of butterflies in the 6-m edges greater than that in the 3-m edges (Fig. 2a). On adjacent ditch banks, too, the number of butterflies (per 100 m length) is virtually the same for the 3- and 6-m wide unsprayed crop edges, and here again there is no additional increase in presence adjacent to a unsprayed 6-m edge (Fig. 2b). Moreover, because of the greater abundance of butterflies on the ditch banks in stead of the crop edges, the absolute butterfly increase will be much higher (per unit area unsprayed crop edge) by creating a unsprayed crop edge of 3 m wide and 100 m long in stead of a unsprayed crop edge of 6 m wide an only 50 m long (see table 2).

Among the other non-target insect groups, flower visitors such as hoverflies (mainly *Episyrphus balteatus*) and aphid predators such as Coccinellidae (*Adalia bipunctata*) were predominant. As can be seen from Figure 3, more insect groups were captured in the 6-m unsprayed edges than in the 3-m unsprayed edges (14.2 versus 11.9). However, in the sprayed situation, too, a difference was found between fields with a 3-m wide edge and those with a

Table 2. Absolute increase of butterflies per 300 m^2 unsprayed crop edge plus adjacent ditch bank, in relation to the sprayed situation, by creating an unsprayed strip of 3 m wide and 100 m long or 6 m wide and 50 m long.

	3 m wide x 100 m long	6 m wide x 50 m long
crop edge	5.4	5.2
adjacent ditch bank	10.9	3.9
total	16.3	9.1

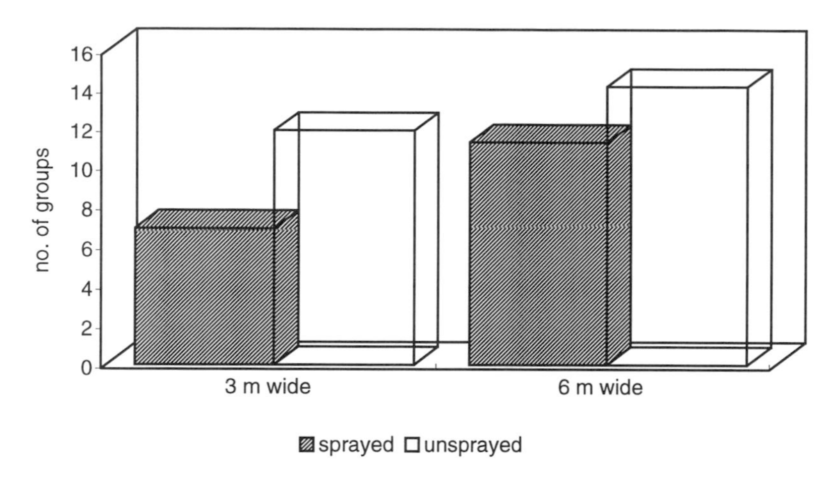

Fig. 3. Comparison of average number of insect groups in 3 and 6 m wide sprayed and unsprayed winter wheat edges (1992). Average no. of insect groups per 100 m, 1.5 m from the field edge.

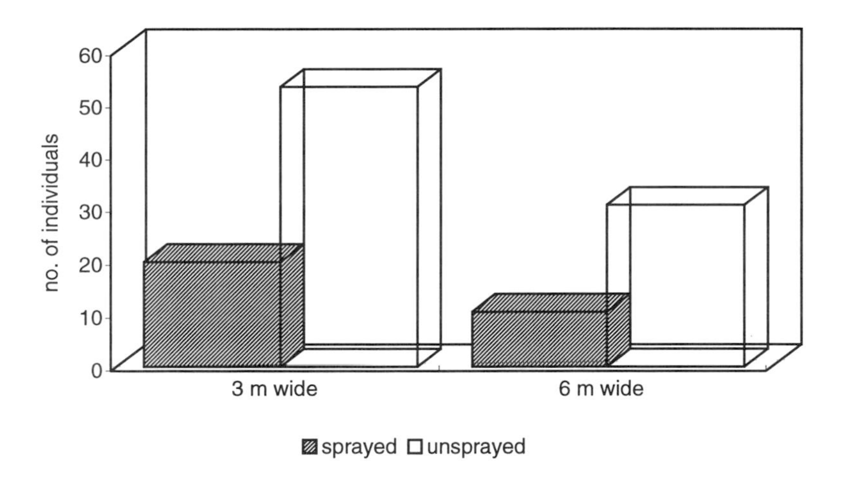

Fig. 4. Comparison of average number of non-target insects in 3 and 6 m wide sprayed and unsprayed winter wheat edges (1992). Average no. of individuals per 100 m, 1.5 m from the field edge.

Table 3. Comparison of dominant non-target insect groups in 3 and 6 m wide sprayed and unsprayed winter wheat edges (1992). Average no. of individuals per 100 m ± standard deviation, 1.5 m from the field edge.

	sprayed fields		unsprayed fields	
	3 m wide	6 m wide	3 m wide	6 m wide
Coccinellidae	5.8 ± 14.2	1.7 ± 3.2	19.0 ± 23.8	4.1 ± 4.8
Chrysomelidae	0.7 ± 0.7	0.8 ± 1.1	4.2 ± 5.5	1.1 ± 1.2
Syrphidae	0.1 ± 0.2	0.4 ± 0.4	1.7 ± 2.7	9.5 ± 22.3
Chrysopidae	3.3 ± 5.3	0.8 ± 1.4	4.7 ± 3.3	2.0 ± 1.5
Heteroptera	0.2 ± 0.3	0.3 ± 0.3	5.0 ± 8.0	1.6 ± 1.8
Heterocera	2.3 ± 1.5	2.8 ± 2.0	4.3 ± 3.9	5.0 ± 3.7

6-m wide edge; the relative increase in the number of insect groups in 6-m edges was no greater than that in 3-m edges (P >0.05, Mann-Whitney U-test). In the unsprayed 6-m edges the average number of individuals is 30.6 per 100 m investigated crop edge and lower than in the 3-m edges (average: 53.2 individuals per 100 m), but this trend was also found in the sprayed situation, and here again the relative increase in the wide strips was no greater than in the narrow strips (Fig. 4). At the level of the individual insect groups, too, the relative increase in insects in the 6-m wide edges was not significantly different from that in the 3-m wide edges (Table 3).

Discussion

As stated already in the introduction, by leaving the outer metres of a crop unsprayed with pesticides, pollution of the surroundings can be drastically reduced. A relatively narrow strip 3 metres wide already gives a very substantial reduction. In the present study, the effects of biodiversity (farmland plants and non-target insects) in the field were investigated. It was found that the highest vegetation cover and most broad-leaved plant species, relevant for flower-visiting insects, are found in the outer 3 metres of the crop edge. Other studies also indicate that the greatest number of species of farmland plants occur in the outer metres of the field (Marshall 1989, Wilson 1989, Smeding & Joenje 1990). If the vegetation cover and number of plant species in unsprayed 3-m crop edges are compared with those in the outer 3 m of 6 m edges, there is found to be no difference. In the 3 m wide edge, therefore, drift deposition does not seem to be a limiting factor. Moreover, in the sprayed situation, too, there is a difference in farmland plant abundance between the inner and outer 3 m. It is thus likely that the greater abundance of farmland plants in the outer 3 m is a natural phenomenon, due to reduced competition from the crop and species exchange with the field boundary. It can therefore be concluded that, with respect to the abundance of farmland plants, a strip 3 metres wide is probably sufficient to achieve maximum number of species and vegetation cover. This is in line with the results of the work by Schumacher (1984), which demonstrates that also in unsprayed crop edges of only 3 metres wide rare species of farmland plants may return.

With respect to the non-target insects, pesticide spraying was found to have a significant impact in both the 3 m and 6 m wide edges. At the same time, however, it was found that by creating unsprayed edges 6 m wide and 450 m long instead of 3 m wide and 100 m long there is no extra enhancement of the number of insect groups or of the density of non-target insects per unit area. So, for this group of organisms, a 3 m unsprayed strip would therefore appear adequate too. Neither does the more isolated location of the short 3 m edges compared with the longer, broader strips appear to play any significant role for the insect groups studied, which are often extremely mobile. However, looking to the adjacent ditch bank, it can be concluded from the butterfly data, that it is preferable to create long unsprayed edges of 3 metre wide in stead of short unsprayed edges of 6 meter wide, when the total unsprayed area is limited. Nevertheless, the results of this study should be interpreted with due caution, since there is major variation in insect presence and abundance among the fields and because the comparison is not entirely sound owing to the fact that the 3 m edges had been left unsprayed for a longer period of time. Insects are unlikely to suffer any time-related effects, though, because for over-wintering they are frequently dependent on the area outside the field.

Naturally, the dimensions of an unsprayed crop edge are not only of relevance for the potential benefits to nature and the environment; due consideration must also be given to the cost of the measures in terms of harvest losses for the farmer. In an unsprayed strip 6 metres wide harvest losses seems to be relatively high in the outer 3 metres of the field probably because of the intensive competition from the farmland plants here. In the inner 3 metres of the strip, more towards the field centre, harvest losses seems to be lower (De Snoo 1994b).

Summarizing, on the basis of the above results it can be stated that from an environmental point of view an unsprayed strip 3 metres wide already yields major benefits to nature and the environment in the vicinity of the field in question. Moreover, the 3 metre wide crop edges appear to offer sufficient perspective for enhancing conditions along the edges for both farmland plants and the insect groups selected for study. If the costs of the measures are also taken into consideration, however, it will be a little more expensive to create a twice as long 3 metre wide strip instead of a shorter 6 metre wide unsprayed strip.

References

Boatman, N.D. 1994. *Field margins: Integrating agriculture and conservation*, BCPC monograph No. 58.

Cuthbertson, P.S. & Jepson, P.C. 1988. Reducing pesticide drift into the hedgerow by inclusion of an unsprayed field margin. *Proc. Brighton Crop Protection Conference, Pests and diseases*: 747-51.

Marshall, E.J.P. 1989. Distribution patterns of plant associated with arable field edges. *J. Applied Ecology* 26: 247-57.

Pollard, E. 1977. A method for assessing changes in the abundance of butterflies. *Biological Conservation* 12, 115-34.

Rands, M.R.W. 1985. Pesticide use on cereals and the survival of grey partridge chicks: a field experiment. *J. Appl. Ecol.* 22: 49-54.

Schumacher, W. 1984. Gefährdete Ackerwildkräuter können auf ungespritzten Feldrändern erhalten werden. *Mitt. LÖLF* 9 (1): 14-20.

Smeding, F.W. & Joenje, W. 1990. Onbespoten en onbemeste perceelsranden in graanakkers. In: Groenendael, J.M., Joenje, W. & Sykora, K.V. (eds.) *10 jaar Zonderwijk & V.P.O. Wageningen*, 129-34.

Snoo, G.R. de. 1994a. Unsprayed field margins in arable land. *Med. Fac. Landbouww. Univ. Gent* 59 (2b): 549-59.

Snoo, G.R. de. 1994b. Cost-benefits of unsprayed crop edges in winter wheat, sugar beet and potatoes. In: Boatman, N.D. (ed.) *Field margins: Integrating agriculture and conservation*, BCPC monograph No. 58: 197-201.

Snoo, G.R. de & Udo de Haes, H.A. 1994. Onbespoten akkerranden voor natuur, milieu en bedrijf. *Landschap* 11 (4): 17-31.

Wilson, P.J. 1989. The distribution of arable weed seedbanks and the implications for the conservation of endangered species and communities. *Proc. Brighton Crop Protection Conference – Weeds 1989*: 1081-86.

Comparison of the carabid fauna of a wheat field and its surrounding habitats

G. Bujaki[1], *Z. Karpati*[1], *F. Kadar*[2], *F. Toth*[1]

[1] Agricultural University, Gödöllô (Hungary)
[2] Plant Protection Institute of Academy of Sciences, Budapest, Hungary

Abstract
To study the connection between the predator populations of cropped and non-crop habitats, the carabid fauna of a winter wheat plot and adjacent habitats was sampled near Gödöllô, Hungary. Pitfall traps were placed in the centre and corners of the wheat field, a grassy field margin and in an area with bushes and trees. From the trap data the following conclusions were drawn:
– Highest numbers of carabids were caught in the adjacent bush area and lowest numbers inside the wheat field.
– Slightly more species were caught in the non-cropped plots. Overall numbers of species were relatively low which may be ascribed to urbanization and isolation effects.
– The faunal composition of the wheat field differed clearly from that in the adjacent habitats. Only three species (*Calathus ambiguus*, *Harpalus distinguendus* and *H. rufipes*) occurred at all sample sites. Of these only *H. rufipes* was abundant in all habitat types and high numbers in non-crop habitats may supplement in field populations.
– There is good evidence that in the area studied, larvae of *H. rufipes, C. ambiguus* and *Broscus cephalotes* overwintered in the wheat plot.

Key words: predator populations, field margins, *Harpalus rufipes, H. distinguendus, Calathus ambiguus, Broscus cephalotes*, overwintering

Introduction

The immediate surroundings of cultivated plant stands (in most cases, grassy field margins or hedgerows) may serve as refuge area under adverse conditions or as breeding place for field-inhabiting arthropods (both pests and beneficial arthropods). These field margins are generally considered to be

Arthropod natural enemies in arable land · II *Survival, reproduction and enhancement*
C.J.H. Booij & L.J.M.F. den Nijs (eds.). *Acta Jutlandica* vol. 71:2 1996, pp. 221-226
 ISBN 87 7288 672 2

advantageous, but their size (Mader and Müller 1984, Welling 1990) and type of plant cover (Welling et al. 1988, Kromp and Steinberger 1992) affect the arthropod composition and hence its function in the agroecosystem. Beneficial arthropods which may be enhanced by the presence of non-crop habitats, are considered to be a major component in the self-regulation of pest in organically grown crops. Organic cultivation itself seems to enhance beneficial arthropod populations in arable fields, as has been proved so far for carabids (e.g. Kromp 1990) and lycosid spiders (Ingrisch et al. 1989). According to Booij & Noorlander (1992) however, the effects of the type of crop present, its structure and crop-related factors on beneficial arthropods outweigh the effects of the farming system applied. The additional presence of suitable non-crop habitats may further increase population levels of beneficials.

In this paper the potential contribution of field surrounding habitats to the carabids in the field is investigated by comparing the fauna of a winter wheat field and its surrounding habitats.

Material and Methods

The surveys have been conducted in the Experimental Farm at the Gödöllô Agricultural University. The area studied was in a winter wheat field (Martonvásári 23 variety) of 4 hectares. The plot was bordered by a dirt road, bushes and a small forest (Fig. 1). On the experimental plot a nitrogen fertilizer (40 kg/hectare) was applied and the field was treated once with a herbicide (Dikotex). No other chemical treatment occurred during the vegetation period. On the plot winter wheat has been grown in three consecutive years.

At six trapping sites (three in the wheat field, and three in surrounding habitats, Fig. 1), the carabid fauna was sampled from mid May to the end of September by three pitfall traps per site. The traps were emptied weekly.

Results

In the course of the surveys in total 980 carabid specimen of 34 different species were caught (Table 1).

Highest numbers of beetles were trapped in the brushy area and the

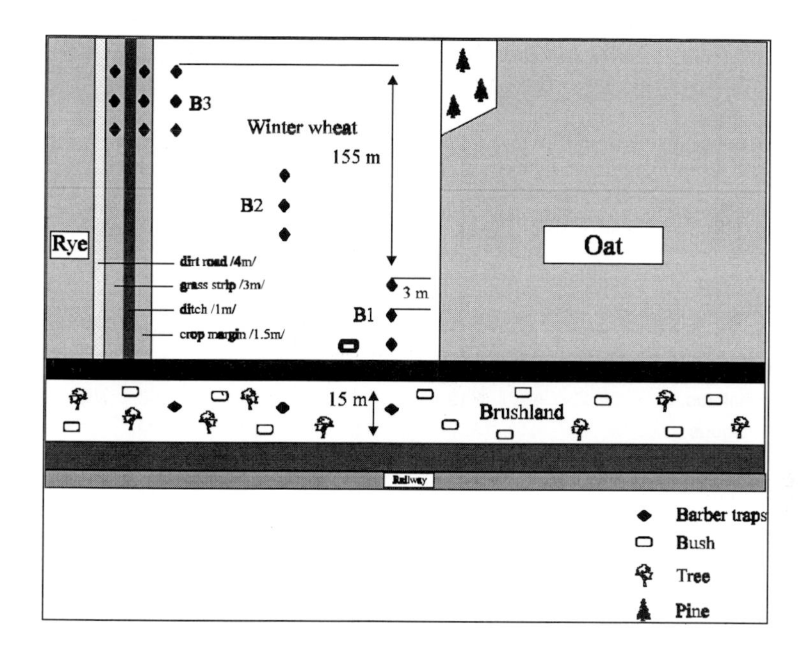

Fig. 1. Schematic map of the investigation area.

lowest in the middle of the wheat stand. Only *Harpalus rufipes* and *Calathus ambiguus* occurred at all sample sites. Relatively high numbers of *Harpalus rufipes* were trapped at the wheat site B1 which is at the field corner close to the brush area where *H. rufipes* is abundant. Similarly *Calathus ambiguus* is frequently trapped in the other field corner B3 near the field margin and grass strip where this species was trapped in considerable numbers. This suggests that for both species some interdependency between habitat patches is apparent.

A total of 17 species were found in the wheat plot of which eight were not caught in the surrounding habitats. Species richness (species/trap) did not differ very strongly between sampling sites, though a few more species were caught in the grass strip and the bushy area. Each habitat appears to have its own characteristic species. For example, the abundant species *Calathus fuscipes* and *Amara bifrons* were almost confined to the non-crop habitats, whereas *Broscus cephalotus* and some other species were exclusively found in the wheat plot, although in low numbers.

Most beetles were captured from the end of July to the end of August. The majority of beetles captured in this period belonged to the species *H. rufipes*. In the non-crop habitats activity extended until the end of September.

Table 1. Carabid activity density in a wheat field and surrounding habitats near Gödöllô, Hungary. Numbers caught from May to September 1994 in three traps/ site.

species	wheat B1	wheat B2	wheat B3	field margin	grass strip	brush strip
Abax parallelopipedus	0	0	0	0	0	3
Amara bifrons	0	0	0	14	4	0
Amara aenea	0	0	0	0	2	0
Amara communis	0	0	0	0	0	1
Agonum dorsale	0	0	0	0	0	1
Bembidion properans	1	2	0	0	0	0
Broscus cephalotus	19	0	1	0	0	0
Calathus ambiguus	7	5	34	79	12	33
Calathus erratus	0	0	0	0	9	0
Calathus fuscipes	0	0	2	13	35	44
Calathus melanocephalus	0	0	0	1	0	0
Calosoma auropunctatus	0	0	1	0	0	0
Carabus coriaceus	0	0	0	1	0	3
Carabus convexus	0	0	0	0	0	4
Carabus violaceus	0	0	0	0	0	1
Dolichus halensis	1	1	0	0	0	0
Drypta dentata	0	0	0	0	0	1
Harpalus affinis	1	1	0	0	0	0
Harpalus atratus	0	0	0	0	0	1
Harpalus azureus	0	0	0	0	2	0
Harpalus cribricollis	0	0	0	1	0	0
Harpalus distinguendus	1	1	0	3	3	1
Harpalus froehlichi	1	0	0	1	0	0
Harpalus rufipes	100	15	20	33	161	252
Harpalus rubripes	0	0	0	0	1	1
Harpalus serripes	0	0	0	0	1	0
Harpalus smaragdinus	1	1	0	0	2	0
Harpalus tardus	0	0	1	0	1	2
Microlestes maurus	0	1	0	0	0	0
Microlestes minutulus	1	3	1	0	1	0
Poecilus cupreus	0	1	0	0	0	0
Poecilus punctulatus	0	3	0	1	0	0
Trechus 4-striatus	0	0	0	0	1	0
Zabrus tenebroides	1	1	2	2	0	0
Carabidae total	134	35	62	149	235	348
Species / trapseries	11	12	8	11	14	14

Discussion

From the results it is clear that the activity-density of carabids in the wheat field was lower than in the surrounding habitats. Also species diversity was relatively low compared to figures given in literature (Booij 1994). The relatively low number of species caught may be connected with the poor sandy soil of the wheat stand, as Kromp & Steinberger (1992) have found much higher species diversity on heavier soils in similar experiments. The lower species diversity may be also due to the environmental circumstances of the wheat stand. It was surrounded by public utilities, so only those species were present that tolerate the effects of urbanization. Impoverishment by isolation may also play a role. The presence of an adjacent railway and a village may limit immigration of species which primarily spread by walking (Wallin 1985).

Although the study area contains many unpaved roads that separate the wheat field from the brush border, their role in preventing immigration seems negligible (Mader et al. 1990). The fact that the overall similarity between the carabid fauna of the wheat field and that of the bordering areas is generally low, seem to be merely due to habitat preferences of different species. Each habitat type has its own characteristic species.

Eight species were established only in the wheat field. Most of them reproduce in autumn and their larvae most likely overwinter in the plot. The carabids *Poecilus cupreus* and *Poecilus punctulatus* occured only in the middle of the plot, showing that winter wheat is a favourable biotope for them. For the six species found exclusively in the brush area, the microclimate, plant density and food supply offered by winter wheat seems to be unsuitable.

The results of this study suggest that the connection between crop and non-crop habitat is weak and limited only to a few species. For *Harpalus rufipes* and *Calathus ambiguus* evidence for connectivity was found. The main distribution of species and individuals may be due to habitat preference. As moving distances within the study area are relatively small compared to the average travelling distances of carabid beetles (Welling 1990), it seems unlikely that potential redistribution in the area is hindered. On the contrary, stronger contrast between habitat patches may be obscured by diffuse movement. Even the distance of 200 m between the furthermost trapping points can be easily covered by a beetle. Especially for low density species the random errors caused by the trapping method is another argument to be made with the interpretation.

References

Booij, C.J.H. & Noorlander, J. 1992. Farming systems and insect predators. *Agricultural Ecosystems & Environment*, 40: 125-35.

Booij K. 1994. Diversity patterns in carabid assemblages in relation to crops and farming systems. In: K. Desender et al. (eds.) *Carabid Beetles: Ecology and Evolution*, 425-31.

Ingrisch, S., Wasner, U. & Glüchk, E., 1989. Vergleichende Untersuchungen der Ackerfauna auf alternativ und konventionell bewirtschafteten Flachen. In: *Alternativer und konventioneller Landbau, Schriftenreihe der Lölf Nordrhein-Westfalen*, Bd. 11: 113-271.

Kiss, J., Kádár, F., Kozma, E., Tóth, I. (1993): Importance of various habitats in agricultural landscape related to integrated pest management: A preliminary study. *Landscape and urban planning*. 27. 191-98.

Kromp, B. & Steinberger, K.H. 1992. Grassy field margins and arthropod diversity: a case study on ground beetles and spiders in eastern Austria (Coleoptera: Carabidae: Arachnida: Aranei, Opiliones). In: Pimentel, D. & Paoletti, M.G. (eds.) *Biotic Diversity in Agroecosystems*. Elsevier, Amsterdam, 71-93.

Kromp, B. 1990. Carabid beetles (Coleoptera, Carabidae) as bioindicators in biological and conventional farmlng in Austrian potato fields. *Biol. Fertil. Soils*, 9: 182-87.

Mader, H.J. & Müller, K. 1984. Der Zusammenhang zwischen Heckenlange und Artenvielfalt. *Z. Kulturtech. Flurbereinig.*, 25: 282-93.

Mader, H.J., Schell, C., Kornacker, P. (1990): Linear barriers to arthropod movements in the landscape. *Biological conservation* 29, 81-96.

Wallin H. (1985): Spatial and temporal distribution of some abundant carabid beetles (Coleoptera: Carabidae) in cereal fields and adjacent habitats. *Pedobiologia* 28, 19-34.

Welling, M. (1990): Dispersal of ground beetles (Coleoptera, Carabidae) in arable land. *Med. Fac. Landbouww. Rijksuniv. Gent*, 55 (2B), 483-91.

Welling, M., Kokta, C., Molthan, J., Ruppert, V., Bathon, H., Klingauf, F., Langenbruch, G.A. & Niemann, P. 1988. Förderung von Nutzinsekten durch Wildkrauter im Feld und im Feldrain als vorbeugende Pflanzenschutzmassnahme. *Schriftenr. Bundesministers für Ernahrung, Landwirtschaft und Forsten (BML), Reihe A: angew. Wiss.*, 365: 56-81.

A review of the progress made to control the cabbage root fly (*Delia radicum*) using parasitoids

S. Finch

Horticulture Research International, Wellesbourne,
Warwick CV35 9EF, UK

Abstract

Some of the classical work on the effects of predatory ground beetles as pest control agents, was done at Wellesbourne in the late 1950s and early 1960s. Much of this work used the cabbage root fly (*Delia radiculum*) as the pest insect and the ground beetles species present in most cultivated fields in northern Europe as the controlling agents. Recent work, however, has shown that these ground beetles are not as effective as predators of the cabbage root fly as previously thought.

In this paper, I will act as the Devil's advocate by questioning: (1) how many of these ground beetles can be considered to be truly "beneficial"?, (2) how many can be relied upon to give predictable levels of control?, and (3) whether the methods now being proposed for enhancing the effects of such beetles are based on sound biological data?

As the regulation of cabbage root fly populations by ground beetles now seems limited, I have now started to work on the two major parasitoids of this pest, one wasp (*Trybliographa rapae*) and one beetle (*Aleochara bilineata*) both of which can infest a relatively high proportion of the overwintering fly pupae. As this subject is new to me, I will use published data to describe my proposed approach. I hope the other participants can highlight the errors of my ways and indicate the lines of research they consider likely to prove most productive.

Key words: beneficial arthropods, enhancement, *Trybliographa rapea, Aleochara bilineata*, competition, conservation

Introduction

During the last 75 years, many authors (e.g. Wadsworth 1915, Tomlin et al. (1992) have suggested that it might be possible to control field populations of

Arthropod natural enemies in arable land · II *Survival, reproduction and enhancement*
C.J.H. Booij & L.J.M.F. den Nijs (eds.). *Acta Jutlandica* vol. 71:2 1996, pp. 227-239
 ISBN 87 7288 672 2

the cabbage root fly (*Delia radicum* L.) using parasitoids. However, no one has yet managed to use parasitoids successfully in this way.

The aim of this review is to determine from the published work which of the methods suggested to date seems most feasible. As I intend this review to be critical rather than encyclopedic, there are likely to be more questions than answers. Nevertheless, by raising the contentious matters, I hope to identify both a profitable future research programme and those areas where additional research is required.

Life-cycle of the parasitoids

Although five species of Braconidae, three of Eucoilidae and four of Ichneumonidae have been reared from cabbage root fly pupae (for authors see Coaker & Finch 1971), the eucoilid *Trybliographa rapae* (Westw.) is the only hymenopterous parasitoid of major importance. This insect lays its eggs in all three larval instars of the fly (Wishart & Monteith 1954) and has been recorded from 60% of the individuals in some samples of overwintering pupae (Wishart et al. 1957). Parasitism of pupae by two beetle species of the genus *Aleochara* is also common. *Aleochara bilineata* (Gyll) is usually more common than *A. bipustulata* (L), possibly because the larvae of *A. bipustulata* find difficulty in entering the puparium of the cabbage root fly, which is thicker than the puparium of its preferred host, the bean seed fly *Delia platura* (Mg.) (Wishart 1957). The two species of *Aleochara* regularly parasitize 20-30% of cabbage root fly pupae (Read 1962) and occasionally 60% (Wishart et al. 1957). The two parasitoid beetles have life-cycles that are out of phase with each other. *A. bipustalata* overwinters as the adult whereas *A. bilineata* overwinters as the first-instar larva within the puparium of its host. Although a few authors have done detailed studies of the life-cycles of the wasp *T. rapae* (Wishart & Monteith 1954) and the beetle *A. bilineata* (Colhoun,1953), most references describe the parasitoids merely as mortality factors in the population dynamics of one, or more, pest species of *Delia*. Therefore, a major aim of the future programme will be to do detailed studies of the life-cycles of the parasitoids to determine how the various species interact, particularly under field conditions.

Proposed ways of using parasitoids in the field

The work to date has centred on *A. bilineata*, mainly because it prefers cabbage root fly pupae to bean seed fly pupae, and because, unlike the wasp *T. rapae*, it is carnivorous throughout its life-cycle. Hence, researchers believe that *A. bilineata* will act both as a predator and parasitoid providing it enters an infested brassica crop sufficiently early in the life-cycle of the pest. Two ways have been suggested for increasing the impact of the two parasitoid beetles. The first is to release *A. bilineata* inundatively at the time the pest fly starts to oviposit. The beetles would then be active much earlier than normal in brassica crops and could first eat the eggs of the fly, to lower the overall pest infestation, and then lay for its progeny to parasitize the remaining pest insects. The second method proposed is more complex (Ahlstrom-Olsson & Jonasson 1992) and is based on using the overwintering adults of the beetle *A. bipustulata* to eat the eggs and early-instars of the fly larvae, and the second beetle *A. bilineata* to eat the later-instars and also parasitize the pupae. This method involves placing considerable amounts of mustard meal around the base of brassica plants to attract and stimulate females of the saprophagous bean seed fly to lay in the mustard meal. The chemicals associated with bean seed fly larvae feeding within this meal then attract the overwintering adults of *A. bipustulata*, which lay in the meal so that their progeny can parasitize the bean seed fly pupae. As such parasitoid beetles are then at the sites where the later-emerging cabbage root fly lay, it is hoped that the beetle adults will feed on cabbage root fly eggs to mature their own eggs, and in this way lower the overall cabbage root fly infestation. The volatile chemicals associated with the feeding of those cabbage root fly larvae that manage to establish on the brassica plants will then attract first the parasitoid wasp *T. rapae,* and later the second beetle parasitoid, *A. bilineata*.

Conserving and "enhancing" the numbers of parasitoids in field crops

Under current farming practices, the major impact of naturally-occurring polyphagous predators and parasitoids is to maintain certain soil-pest populations at more or less constant levels from year to year. As insecticides all act in a density-independent manner, large fluctuations in pest populations cannot be tolerated if reduced application rates of insecticide are to remain consistently effective in future (Suett & Thompson 1985). Therefore, our principle aim at present is to "conserve" the existing levels of natural pest control. Unfortunately, in most cases, there is little quantitative information on the contribution made to overall pest control by the various biological agents. What is known is that parasitism can be either high or low in pest populations. What is not known is whether we can manipulate the field environment to ensure that parasitism is high in all instances. Many researchers within the IOBC Working Group "Integrated Farming Systems" (see Vereijken & Royle 1989) are now attempting to increase populations of predators and parasitoids by the introduction of additional feeding sites and refuges. The problem with this approach is in assuming that the more beneficial insects there are in the area, the better the control of pest species will be. Pest control may not improve within such systems, particularly if the increase in the numbers of predators/parasitoids merely reflects the increase in the numbers of alternative sources of prey/hosts within the new habitats. Although it might seem like semantics, "increasing" and "enhancing" parasitoid numbers are not synonymous. The word enhance means "to add to the effect". In the field, it is relatively easy to "increase" parasitoid numbers whereas "enhancing" their effects is much more difficult. For example, parasitoid numbers can be "increased" by releasing parasitoids into fly-infested crops. However, if such parasitoids disperse before laying, then obviously they will not "enhance" the levels of pest control locally.

Changing the environment and its effects on parasitoids of the cabbage root fly

Flowering plants

Improving the general environment for predatory insects and parasitoids is attempted usually by growing flowering plants in or around monocultures, to provide the beneficial insects with additional sources of food in the form of nectar and/or pollen (Finch 1988). This approach might help to increase the fecundity of *T. rapae*, once the types of flowers visited have been identified. This approach will not directly improve the environment for either species of *Aleochara*, however, as they feed almost entirely on animal protein.

Grassy banks

Attempts are being made to introduce refuges, such as grassy banks, into crop fields, to increase the numbers of predators and parasitoids that overwinter successfully (Thomas et al. 1991). This approach is unlikely to be effective with *T. rapae* or *A. bilineata* as both overwinter within cultivated fields inside the puparia of their host flies. Refuges might be effective, however, for *A. bipustulata*, particularly if brassica crops are grown after hay (*Lolium perenne* L.), as the plant material left once hay crops have been harvested can support large numbers of bean seed flies and could leave large numbers of *A. bipustulata* overwintering in such fields. The main problem with this approach is that there is no information on whether the refuges increase overall survival or merely alter the distribution of the overwintering beetles.

Undersown crops

The final way to make the environment more diverse within cultivated crops is to undersow the main crop with a forage crop, such as clover (*Trifolium repens* L.). Such systems have an adverse effect on pest insects largely because the pest insects are adapted to plants growing in bare soil (Kostal & Finch 1994). Unfortunately, the beetle parasitoids of the cabbage root fly are also strongly adapted to bare soil situations and so undersowing with clover reduces the effectiveness of *A. bilineata*, though it improves slightly the levels of parasitization by *T. rapae* (Langer 1992).

Problems to be considered when releasing parasitoids of the cabbage root fly

Short-season crops

Three factors operate against the use of "classical" biological control in brassica crops: the short growing season of many crops, the transience of the crops due to rotation, and the demand for increasingly high-quality produce. These three factors are complementary as, in the short-term relationship between pest and crop, the natural enemies of the pest have insufficient time to establish their superiority before the crop is damaged and no longer of high quality. This is particularly true for the cabbage root fly, as this fly generally enters brassica crops shortly after they have been transplanted and before the transplants have had time to establish adequate root systems. Therefore, as most brassica crops need to be protected more or less as soon as they are planted, the parasitoids will have to be released shortly after the crop is planted if they are to be effective as predators.

Rearing sufficient parasitoids

The main difficulty in mass-rearing parasitoids of the cabbage root fly is that an artificial diet has not yet been developed and so the host insect still has to be reared on swedes (*Brassica napus* var. *napobrassica*) potted in sand (Finch & Coaker 1969) which is both labour-intensive and physically-demanding. There is always the possibility of rearing the parasitoids on the closely-related onion fly (*Delia antiqua*), for which there is an artificial diet (Ticheler et al. 1980), but getting them to "switch" back onto the cabbage root fly might then become a problem. Perhaps the percentage of beetles that can "switch" from developing on onion fly pupae to developing on cabbage root fly pupae can be maintained at a high level by regularly transferring the insect culture back onto cabbage root fly pupae after having reared them on onion fly pupae for a fixed number of generations? However, it is possible that the beetles may have to be reared solely on cabbage root fly, as there are indications that they may not switch readily. For example, in recent experiments females of *A. bilineata* provided with third-instar cabbage root fly larvae as food started to lay high numbers of eggs after three days. In contrast, similar beetles provided with second-instar onion fly larvae as food laid only a few eggs after 21 days. The latter beetles, however, started to produce high numbers of eggs as soon as they were presented with third-instar cabbage root fly larvae (Finch & Vreugdenhil - unpublished data). It is not yet clear whether this resulted from a difference between the two fly species or merely because of the different size of the prey items provided as food.

Cost of crop protection

This will depend upon the number of staphylinid beetles required. The only two estimates made at present vary between 20,000 (Bromand 1980) and 650,000 (Hertveldt et al. 1984) beetles per hectare. Using the rearing technique described by Whistlecraft et al. (1985), ten hours of labour would be needed to produce 20,000 *A. bilineata.* Although the work required for this approach might seem daunting, the costs appear to compare favourably with insecticidal control. For example to treat 20,000 plants (1 ha) with chlorfenvinphos granules requires 11.2 kg of product at a cost of £4.63/kg, or about £52/ha. Therefore, providing the hourly wage, plus overheads, of the workers producing the beetles does not exceed £5, the cost of control using these parasitoid beetles could be similar to that of using insecticide. Application costs might vary, but as chlorfenvinphos granules are applied generally as a sub-surface band at a cost of about £30/ha, there appears to be sufficient flexibility to keep within this cost even if the beetles have to be released manually.

Distributing the insects within the crop to be protected

According to Esbjerg & Bromand (1977), *A. bilineata* released into brassica crops disperse at the rate of about 6.5 m per day. From studies with beetles marked with radioisotopes, they concluded that for the control of cabbage root fly, batches of several hundred beetles should be placed at each release point, which should be spaced no more than 20 m apart to ensure that the beetles spread throughout the crop as quickly as possible. Based on such data, releasing beetles from 16-20 points/ha should not create problems. To be effective, it is likely that the beetles will have to be distributed in this way, as females of the cabbage root fly are distributed more or less evenly through brassica crops (Finch & Skinner 1973).

Competition with other parasitoids

Although many people believe that an array of parasitoids is preferable to using just one species, when the parasitoids compete for the same resource, the presence of more than one species can prevent the build-up of large parasitoid populations. For example, although the wasp *T. rapae* infests all three of the larval instars of the fly, the wasp larva does not develop past the first-instar phase until the fly larva pupates. Despite the small size of this first-instar parasitoid larva, it has a profound effect on the development of its host, as the wasp larva must feed from the fly larva to keep itself alive.

Consequently, the earlier a fly larva is attacked by a wasp the smaller the subsequent fly pupa. Once the fly larva has pupated it can be parasitized also by *A. bilineata*. Reader & Jones (1990) showed that there were marked differences in both the species emerging and in overall mortality when pupae were attacked by larvae of both parasitoids. In these multiparasitized pupae, insects emerged from only 13% of the pupae, when the wasp larva was still feeding within the fly pupa, and all of the insects were *A. bilineata.* In contrast, when the wasp larva had moved out of the fly pupa and was feeding on it from the outside inwards, insects emerged from 65% of the multiparasitized pupae and all of them were wasps. Hence, depending upon when the second parasitoid attacks, the mortality of the competing parasitoids can vary between 35% and 87%. The proposed 3x3 ("9 cell") parasitoid matrix (Fig. 1), showing three rows of pupae to indicate early, intermediate and late attacks by the beetle, can be used to describe how the wasp and beetle are probably interacting. Whether or not a parasitoid emerges from a multiparasitized pupa is governed both by the time the beetle larva attacks the fly pupa and by the size of the pupa. The latter is governed largely by how long the wasp larva has been inside the fly larva. Attacks by the wasp on first-, second- and third-instar fly larvae result in the production of small, medium and large (normal) fly pupae. When the beetle larva attacks a small pupa containing a wasp larva which is still feeding inside the fly pupa (cell 1), both insects die as there is insufficient food available for either insect to complete its development. In contrast, once the wasp larva has moved to feed externally it should outcompete the beetle (cell 7), though undoubtedly if it has only just started feeding in this manner (cell 4), neither may gain sufficient food to complete development. The same could be true for the medium sized pupae (cell 2) in which the wasp larva is still within the fly pupa and hence once a beetle starts to feed there may not be sufficient food for either the poorly developed wasp larva or the beetle larva to survive. With the exception of "cell 3", wasps should emerge from all of the other multiparasitized pupae (cells 5-9). The type of pupae found in cell 3 are the only ones that still contain sufficient reserves to allow the beetle larva both to outcompete the wasp larva and to complete its development. Nevertheless, to be successful the beetles must enter such pupae early in their development. The competition between the two parasitoids is biased heavily in favour of the wasp, as even if the beetle larva finds a cabbage root fly larva shortly after it has pupated, the beetle larva still needs between 12 and 36 hours to chew its way through the wall of the puparium before it can start feeding on the fly pupa (Colhoun 1953). The percentage of the total individuals that each cell represents depends on the proportion of fly larvae parasitised in each instar. Hence, it might be assumed from Fig. 1 that cell 3 represents 33% of the *T. rapae* larvae still feeding within the fly pupa. However, the actual percentage is much lower than this because the wasps parasitize mainly the two earlier instars.

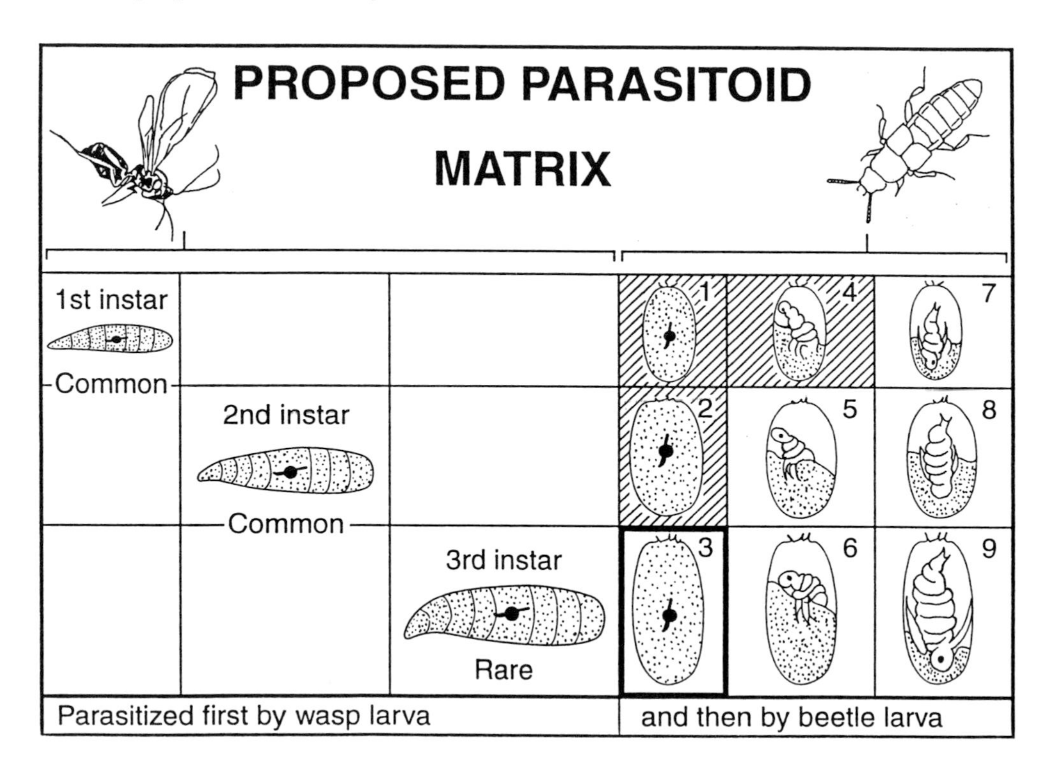

Fig. 1. Proposed survival of insects from cabbage root fly pupae multiparasitized by the eucoilid wasp *Trybliographae rapae* and the staphylinid beetle *Aleochara bilineata.* The wasp can attack each of the three larval instars of its fly host and this governs whether the resulting fly pupae are small, medium, or large. The effect that the beetle larvae have depends upon whether they enter when the wasp larva is young (cells 1-3), of intermediate age (cells 4-6), or well-developed (cells 7-9). Matrix developed mainly from the findings of Reader & Jones (1990).

The above information shows that when parasitoids compete, many of their progeny die from multiparasitism. The relationship between the percentage of flies (y) and the percentage of parasitoids (x) that emerged from cabbage root fly pupae collected from ten sites in England and Wales was $y = 86-1.6x$ (Finch

& Collier 1984), indicating that no flies emerged when 55% of the pupae gave rise to living parasitoids.

Competition with other natural enemies
One of the problems of releasing any biological agent into the field is that it has to compete with the established natural enemies. At worst, the release may simply upset the overall local balance so that the existing predators feed in a density-dependent manner on the released parasitoids until the balance is re-established. At such times, it is questionable whether polyphagous predatory beetles should be regarded as "beneficial" insects.

Effects of cultural practices on parasitoid numbers
Changes in the level of parasitism at a particular site can occur either gradually or rapidly. For example, at one site in Denmark, Bromand (1980) noted a gradual decline in the parasitism of overwintering cabbage root fly pupae by *A. bilineata* from 1971 to 1975 and associated it with a decrease in the area of swedes being grown. He suggested that parasitism had declined because, as fewer fields were planted, a larger proportion of the beetles failed to find brassica crops. In contrast, rapid reductions in the levels of parasitism can be caused by some of the soil insecticides applied regularly to control the cabbage root fly (Coaker 1966, Bromand 1980), the parasitoids being more susceptible to these insecticides than the host insect (El Titi 1980, Finch & Skinner 1980).

Apart from the adverse effects of pesticides on parasitoids, studies are needed on why ploughing has a greater effect on parasitoids than on pest populations (Finch & Skinner 1980); why parasitized pupae are rarely recovered from crops grown in highly-organic soils (unpublished data); and how cultural practices such as intercropping and applying fertilizers, particularly organic ones, affect both pest and parasitoid numbers (Theunissen & den Ouden 1980).

Discussion

It is clear from this review that trying to enhance the activity of any one parasitoid within a group complex is not easy, as changes to "improve" the environment for one species invariably have the opposite effect on one of the other species. For example, growing flowering plants alongside crop boundaries could provide additional feeding sites for the wasp. However, such plants would undoubtedly support additional prey species that could lower the impact of the beetles as predators of the cabbage root fly (Finch 1988).

The main problem with all of the systems used until now to "enhance" the numbers of polyphagous predators and parasitoids, is that the various treatments have only added to the numbers of beneficial insects present rather than "enhancing" their effects as pest control agents. This raises the question of whether it will ever be possible to improve pest control by making crop boundaries more diverse, as the associations between the numbers of predators and their prey and the numbers of parasitoids and their hosts, is finely balanced in such systems. In general, insects only attain pest status in the types of unbalanced systems that occur in agriculture and, in particular, in large monocultures. Therefore, perhaps the only way to resolve this problem will be to treat the unbalanced systems with an unbalanced control measure, such as the release of higher than normal numbers of laboratory-reared predators/parasitoids. At present we use specific insecticides to control the cabbage root fly, so by analogy we may also need to use specific, rather than general, biological agents for the types of pest control needed in ephemeral cultivated crops.

From this review, it appears that it should be possible to use one of the three common parasitoids to control the cabbage root fly. Unfortunately, the cabbage root fly is not the ideal insect on which to mass-rear parasitoids, as the fly cannot be reared on an artificial diet. However, despite the work involved, it might be prudent to test the feasibility of this approach by rearing the parasitoids in cabbage root fly larvae/pupae, as the extremely high numbers of insects needed by Hertveldt et al. (1984) to produce an effect against this species in the field, suggests that parasitoid beetles (*A. bilineata*) reared on onion fly may not "switch" too readily onto cabbage root fly in the field.

Many authors have indicated that the beetle *A. bilineata* is probably the most appropriate parasitoid to rear and release against the cabbage root fly. However, the published data indicate that this would be true only in localities where the wasp *T. rapae* does not occur. Although such localities can be found in Canada (Wishart & Monteith 1954), the wasp was found to be the dominant species in ten countries in northern Europe (Finch et al. 1985). A second drawback to releasing the beetle *A. bilineata* as a predator, is that it would have to be released at the start of each and every fly generation as, being a pupal parasitoid, no matter how early it is released its offspring will always emerge 2-3 weeks later than the pest fly.

The second alternative is to release the wasp, but as this is not predatory, the benefits of high levels of parasitism would accrue only in subsequent generations. There does, however, seem to be considerable scope for increasing levels of parasitization by the wasp, as no beetles emerged from samples of cabbage root fly pupae collected from Belgium, Denmark, Eire, Germany, Northern Ireland and The Netherlands, and in many of these countries the levels of parasitization by the wasp rarely exceeded 5% (unpublished data). Therefore, in these countries, competition between the two major parasitoids appeared to be of little importance. Presumably the best way to reduce the competition from the beetle *A. bilineata* in a specific locality in the UK would be to release wasps early into a crop so they parasitize a higher proportion of the early-instars of the fly and hence reduce considerably those instances of multiparasitism where the beetle larva was likely to survive. In addition, if *T. rapae* is to be reared in the laboratory, it would help if the fly larvae could be reared on an artificial diet to make them more accessible to the wasps. Alternatively, high levels of parasitism may be easy to achieve if the wasps will attack first-instar larvae shortly after they emerge from the eggs.

The third parasitoid species, *A. bipustulata*, has the advantage that it overwinters as the adult and hence is active at the cooler temperatures that occur in the field in early spring. If it could be established in high numbers, then this species might be able to maintain such populations naturally. Like the other two species, however, *A. bipustulata* is not without its problems. For example, if reasonable numbers of beetles are to be produced they may have to be reared on the bean seed fly, which could create problems with "switching". Furthermore, it is assumed generally that this parasitoid will eat cabbage root fly eggs under field conditions but, as yet, no data have been collected to this effect.

Whether, or not, *A. bipustulata* can be used in conjunction with the wasp to give adequate levels of control requires further testing. What is certain, however, is that if attempts are made to arrest this beetle around the base of brassica plants by adding decomposing organic material, something other than mustard-meal (Ahlstrom-Olsson & Jonasson 1992) should be used, as when Ahlstrom-Olsson tested the technique in a swede crop at Horticulture Research International Wellesbourne in 1992, the mustard meal attracted preferentially the cabbage root fly and hence the plants surrounded by mustard meal were damaged more severely than the plants without mustard meal.

Acknowledgement

I wish to thank the MAFF for supporting this work as part of Project HH1815SFV.

References

Ahlström-Olsson, M. & Jonasson, T. 1992. Mustard meal mulch – a possible cultural method for attracting natural enemies of brassica root flies into brassica crops. *OILB/SROP Bulletin* 15/4: 171-75.

Bromand, B. 1980. Investigations on the biological control of the cabbage root fly (*Hylemya brassicae*) with *Aleochara bilineata. Bulletin OILB/SROP* 3/1: 49-62.

Coaker, T.H. 1966. The effect of soil insecticides on the predators and parasites of the cabbage root fly (*Erioischia brassicae* (Bouché)) and on the subsequent damage caused by the pest. *Annals of Applied Biology* 57: 397-407.

Coaker, T.H. & Finch, S. 1971. The cabbage root fly, *Erioischia brassicae* (Bouché). *Report of the National Vegetable Research Station for 1970:* 23-42.

Colhoun, E.H. 1953. Notes on the stages and the biology of *Baryodma ontarionis* Casey (Coleoptera: Staphylinidae), a parasite of the cabbage maggot, *Hylemya brassicae* Bouché (Diptera: Anthomyiidae). *Canadian Entomologist* 85: 1-8.

El Titi, A. 1980. Die Veränderung der Kohlfliegenmortalitat als Folge der chemischen Bekämpfung von anderen Kohlschadlingen. *Zeitschrift fur angewandte Entomologie* 90: 401-12.

Esbjerg, P. & Bromand, B. 1977. Labelling with radioisotopes, release and dispersal of the rove beetle, *Aleochara bilineata* Gyll. (Coleoptera: Staphylinidae) in a Danish cauliflower field. *Tidsskrift for Planteavl* 81: 457-68.

Finch, S. 1988. Entomology of crucifers and agriculture. Diversification of the agroecosystem in relation to cruciferous crops. In: Harris, M.K. & Rogers, C.E. (eds.) *The Entomology of Indigenous and Naturalized Systems in Agriculture.* Westview Press: Boulder Colorado, 39-71.

Finch, S. & Coaker, T.H. 1969. A method for the continuous rearing of the cabbage root fly *Erioischia brassicae* (Bouché) and some observations on its biology. *Bulletin of Entomological Research* 58: 619-27.

Finch, S. & Collier, R.H. 1984. Parasitism of overwintering pupae of cabbage root fly, *Delia radicum* (L.) (Diptera: Anthomyiidae), in England and Wales. *Bulletin of Entomological Research* 74: 79-86.

Finch, S. & Skinner, G. 1973. Distribution of cabbage root flies in brassica crops. *Annals of Applied Biology* 75: 1-14.

Finch, S. & Skinner, G. 1980. Mortality of overwintering pupae of the cabbage root fly [*Delia brassicae*]. *Journal of Applied Ecology* 17: 657-65.

Finch, S., Bromand, B., Brunel, E., Bues, M., Collier, R.H., Foster, G., Freuler, J., Hommes, M., Van Keymeulen, M., Mowat, D.J., Pelerents, C., Skinner, G., Stadler, E. & Theunissen, J. 1985. Emergence of cabbage root flies from puparia collected throughout northern Europe. In: Cavalloro, R. & Pelerents, C. (eds.) *Progress on Pest Management in Field Vegetables.* P.P. Rotondo – D.G. XIII – Luxembourg No. EUR 10514. Balkema Rotterdam, 33-36.

Hertveldt, L., Van Keymeulen, M. & Pelerents, C. 1984. Large scale rearing of the entomophagous rove beetle *Aleochara bilineata* (Coleoptera: Staphylinidae). *Mitteilungen aus der Biologischen Bundesanstalt für Land- und Forstwirtschaft* 218: 70-75.

Kostal, V. & Finch, S. 1994. Influence of background on host-plant selection and subsequent oviposition by the cabbage root fly (*Delia radicum*). *Entomologia experimentalis et applicata* 70: 153-63.

Langer, V. 1992. The use of a living mulch of white clover on the control of the cabbage root fly (*Delia radicum*) in white cabbage. *IOBC/WPRS Bulletin* 15/4: 102-3.

Read, D.C. 1962. Notes on the life history of *Aleochara bilineata* (Gyll.) (Coleoptera: Staphylinidae), and its potential value as a control agent for the cabbage maggot, *Hylemya brassicae* (Bouché) (Diptera: Anthomyiidae). *Canadian Entomologist* 94: 417-24.

Reader, P.M. & Jones, T.H. 1990. Interactions between an eucoilid [Hymenoptera] and a staphylinid [Coleoptera] parasitoid of the cabbage root fly. *Entomophaga* 35: 241-46.

Suett, D.L. & Thompson, A.R. 1985. The development of localised insecticide placement methods in soil. *British Crop Protection Council Monograph* 28: 65-74.

Theunissen, J. & den Ouden, H. (1980). Effects of intercropping with *Spergula arrensis* on pests of Brussels sprouts. *Entomologia experimentalis et applicata* 27: 260-68.

Thomas, M.B., Wratten S.D. & Sotherton, N.W. (1991). Creation of "island" habitats in farmland to manipulate populations of beneficial anthropods. *Journal of Applied Ecology* 28: 906-17.

Ticheler, J., Loosjes, M. & Noorlander, J. 1980. Sterile-insect technique for control of the onion maggot, *Delia antiqua*. In: *Integrated Control of Insect Pests in the Netherlands*, Pudoc, Wageningen, 93-97.

Tomlin, A.D., McLeod, D.G.R., Moore, L.V., Whistlecraft, J.W., Miller, J.J. & Tolman, J.H. 1992. Dispersal of *Aleochara bilineata* [Col.: Staphylinidae] following inundative releases in urban gardens. *Entomophaga* 37: 55-63.

Vereijken, P. & Royle, D.J. 1989. Editors of : Current Status of Integrated Farming Systems Research in Western Europe. *IOBC/WPRS Bulletin* 12/5: 76 pp.

Wadsworth, J.T. 1915. On the life-history of *Aleochara bilineata* Gyll., a staphylinid parasite of *Chortophila brassicae*, Bouché. *Journal of Economic Biology* 10: 1-27.

Whistlecraft, J.W., Harris, C.R., Tolman, J.H. & Tomlin, A.D. 1985. Mass-rearing technique for *Aleochara bilineata* (Coleoptera: Staphylinidae). *Journal of Economic Entomology* 78: 995-97.

Wishart, G. 1957. Surveys of parasites of *Hylemya* spp. (Diptera: Anthomyiidae) that attack cruciferous crops in Canada. *Canadian Entomologist* 89: 450-54.

Wishart, G. & Monteith, E. 1954. *Trybliographa rapae* (Westw.) (Hymenoptera: Cynipidae), a parasite of *Hylemya* spp. (Diptera: Anthomyiidae). *Canadian Entomologist* 86: 145-54.

Wishart, G., Colhoun, E.H. & Monteith, E. 1957. Parasites of *Hylemya* sp. (Diptera: Anthomyiidae) that attack cruciferous crops in Europe. *Canadian Entomologist* 89: 510-17.

Can sustainable agricultural practices affect biodiversity in agricultural landscapes? A case study concerning orchards in Italy

Maurizio G. Paoletti, D. Sommaggio, M. Bressan & V. Celano

Department of Biology, Padova University,
Via Trieste, 75-35121 Padova, Italy

Abstract
Up to 95% of the terrestrial environment is affected by human activities, with agriculture in particular producing the greatest modifications. The loss of biodiversity in agroecosystems has been described by many authors. Our research studies the changes in macroinvertebrates in various orchards subjected to different agricultural practices. Consistent data shows that biologically farmed landscapes support a higher number of biota. Therefore we believe that organisms such as Isopoda, Lumbricidae, Araneae and Carabidae, sensitive to agricultural practices, may be used as bioindicators of agricultural sustainability.

Key words: bioindicators, invertebrates, sustainability indicators, carabidae, earthworms, orchards

Introduction

"Biodiversity conservation entails a shift from a defensive posture – protecting nature from the impacts of development – to an offensive effort seeking to meet people's needs from biological resources, while ensuring the long-term sustainability of the Earth" (Raven 1992).

The basic DNA information occurs in an estimated 1.4-1.8 million currently known species, dominated by insects (Wilson 1988, Stork 1988, Wheeler 1990, May 1992). The total number of species is estimated at 7.3-100 million (Wilson 1988, Erwin 1988, Ehrlich and Wilson 1991, May 1992). Therefore the species on our planet are estimated to be 5-50 times more abundant than those at present identified and described. More recently Ehrlich and Wilson (1991) have estimated that living species could reach the 100 million mark!

Arthropod natural enemies in arable land · II *Survival, reproduction and enhancement*
C.J.H. Booij & L.J.M.F. den Nijs (eds.). *Acta Jutlandica* vol. 71:2 1996, pp. 241-254
 ISBN 87 7288 672 2

There are at least two points which amaze the researcher. The first is the number of insect species we have on the planet and the second is how few plant and animal species we currently consider suitable as food. For example, in Italy out of 5300 plant species, those cultivated are no more than 100, but the really important ones are very few, only 11. A similar situation can be found in many other Western countries.

If agriculture is an activity based on only a few annual plants and a limited number of animals, it is nevertheless still the primary activity affecting the largest areas in most countries (covering from 50 to 90% of their territory). In spite of this, a consistent number of organisms are directly or indirectly linked with agricultural landscapes: more than those considered by agriculturists.

Due to their response to agricultural practices, soil invertebrates may help in "reading" landscape stress and evaluating the more sustainable (less disruptive) agricultural practices. In this paper we will show how invertebrate biodiversity may be used as indicators of sustainability in orchard agroecosystems.

Materials and Methods

Forli, Italy: assessment of invertebrates in six peach orchards

To monitor invertebrates as a means of evaluating sustainability, we selected six farms for the first two-year trial. The farms were divided into three groups: biological farms (B), integrated pest management farms (IPM), conventional farms (C). The three groups were monitored initially to ensure they were of similar dimension, location and soil type, but had different inputs which increased from low in the biological orchard to high in conventional ones. The two biological orchards, B1 and B2, had a permanent weed cover between the rows of peach trees. The two integrated farms differed in soil tillage: IPM1 had a permanent weed cover, whereas in IPM2 the soil was tilled each month from April to September to control weed growth. Similarly, both of the conventional orchards were tilled periodically to destroy weeds between the peach trees. Weed cover was similar in B1, B2 and IPM1, with a prevalence of *Trifolium*.

Soil invertebrates were sampled each month (from March to November, 1991, and from April to November, 1992) using six pitfall traps (7 cm diameter) per farm. Traps were filled with a solution of propylene glycol:water (30:70) and left open in the field for 20 days. Pitfall traps were placed in the rows between peach trees, 7-10 days after soil tillage. Data are expressed as activity density (Thiele 1977, Brandmayr and Brunello-Zanitti 1982).

Most of the taxa were identified according to species (Paoletti et al. 1992). During first sampling invertebrates above the vegetation were also sampled; data on this subject together with data on soil mesofauna collected with a Tullgren modified extractor have been published in another paper (Paoletti et al. 1992).

Assessment of earthworms in 64 orchards in Forlì

In 1993, earthworm populations were monitored in 64 orchards. The selected orchards belong to four different types, characterised by different crops but also by different chemical inputs (16 orchards of each type):
– Apple orchards: high chemical input
– Vineyards: medium chemical input but with high input of copper (resulting from $CuSO_4$ applied as main fungicide)
– Peach orchards: medium chemical input and with low input of copper (resulting from $CuSO_4$ applied as a fungicide),
– Kiwi fruit orchards: low chemical and low input of copper (resulting from $CuSO_4$ applied as a fungicide).

In each type of orchard different soil tillage management was practised: eight orchards were tilled and eight were covered with vegetation.
From each orchard we collected five 25x25 cm, 20 cm deep soil samples, in April and October 1993, after a heavy rainfall. The earthworms, collected by hand sorting, were preserved in 75% alcohol, and later on identified and weighed. In April, the soil from four sampling sites in 32 of the orchards (eight of each type: four tilled and four untilled) were analysed for Cu, Zn (with atomic absorption spectrometer and N, P, Ca, Mg, K, Na and organic matter (S.I.S.S. 1985).

Results

The total abundance of isopods was reduced in integrated and conventional orchards (Fig. 1) and the activity density in integrated and conventional orchards was only 3-19% of that in biological ones. Araneae activity density was higher in biological orchards, especially B2. The distribution of ground beetles in the six orchards does not appear to be linked to agricultural inputs, as we observed higher activity in the integrated orchards, and similar numbers in both the high and low input fields. In 1991, IPM1 had the highest number of specimens, in 1992 IPM2 presented the highest activity density.

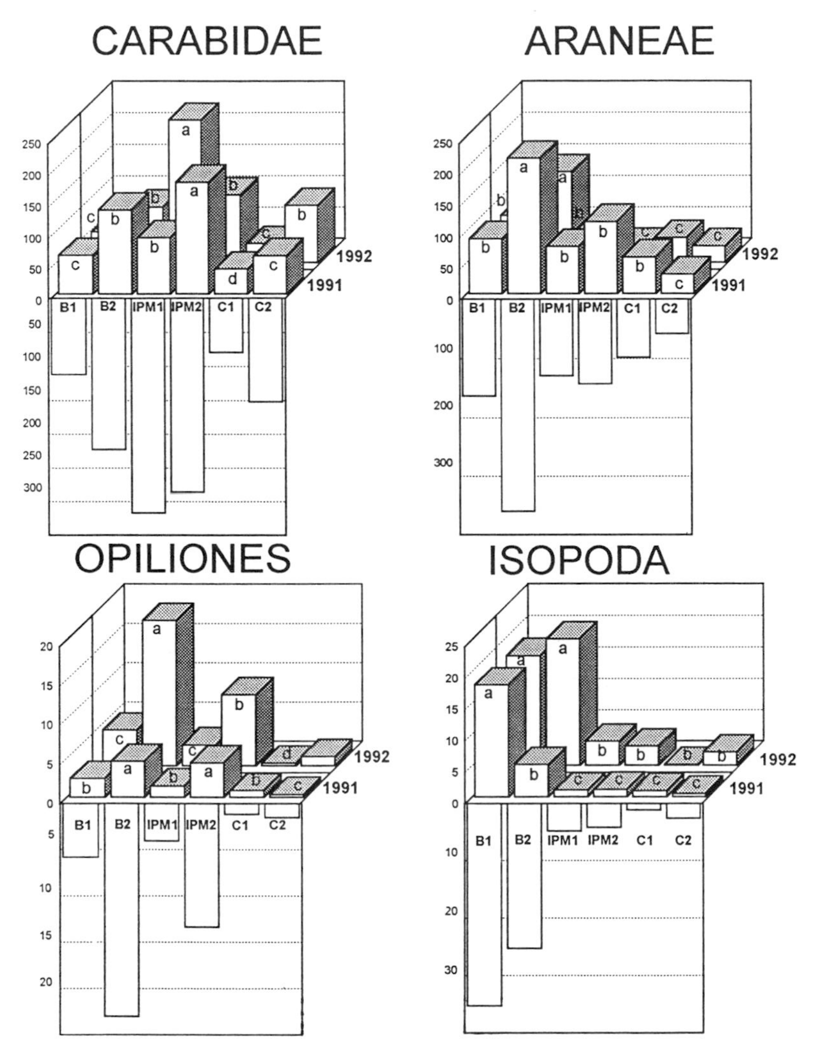

Fig. 1. Total activity density of various invertebrates caught in pitfall traps. Different letters represent statistically different numbers (Kruskall-Wallis test, $P < 0.05$). Values are the total of ten sampling data in 1991 and eleven in 1992. The low part of the graph represents the sum of activity density of the two years. Legend: B1 and B2: organic orchards; IPM1: integrated pest orchard with a spontaneous cover vegetation; IPM2: integrated pest management orchard with soil tillage; C1 and C2: conventional orchards.

Biodiversity in orchards with different inputs

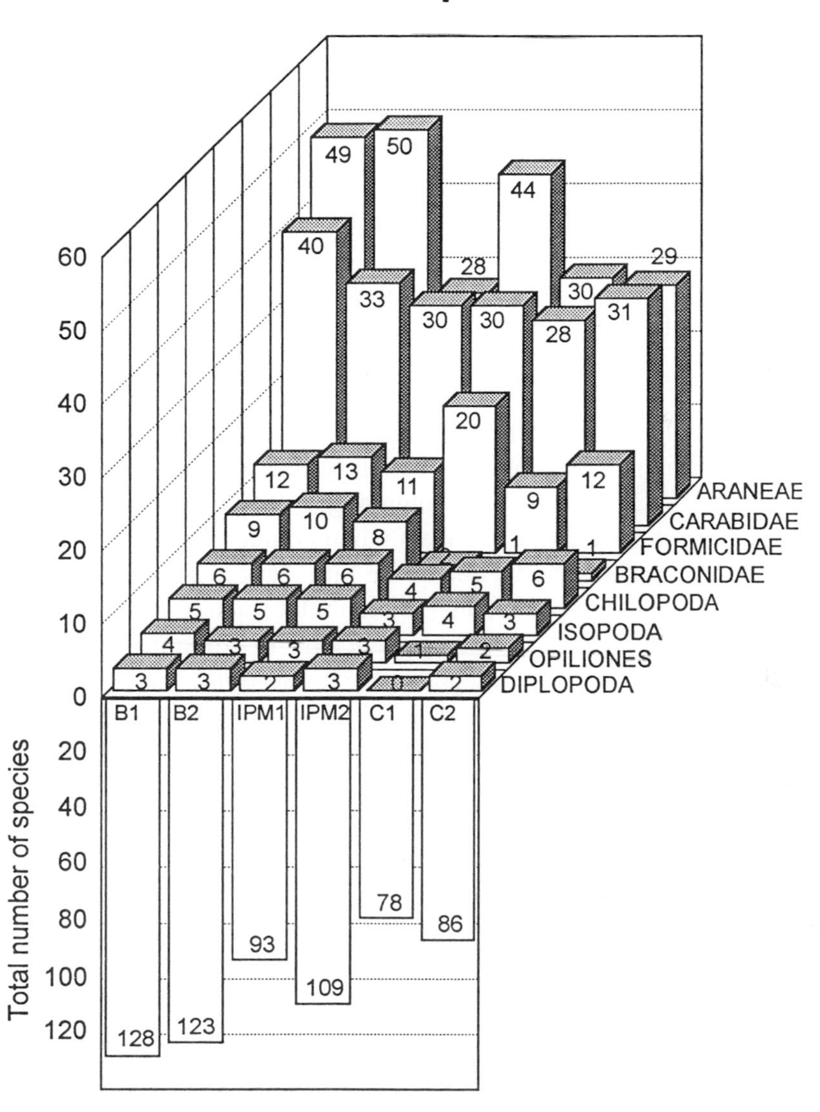

Fig. 2. Total number of species for all taxa classifed at species level. Legend: B1 and B2: organic orchards; IPM1: integrated pest orchard with a spontaneous cover vegetation; IPM2: integrated pest management orchard with soil tillage; C1 and C2: conventional orchards.

However, the high number of carabid beetles collected from the integrated orchards is not correlated with an equally high number of species (see Fig. 2). The high number of specimens is explained by the presence of species such as *Pseudophonus rufipes* De Geer, *Bembidion properans* Steph. and *Harpalus distinguendus* Dft., typical of disturbed habitats. For example, regarding the high number of ground beetle specimens collected in IPM1 during 1992, 71.8% consists of *Ps. rufipes*.

We also observed an almost 40% loss of Araneae species in the untilled integrated and conventional orchards (Fig. 2). Considering all taxa classified at species level, 30-39% fewer species were found in conventional orchards than in organic orchards. In integrated orchards the loss of biodiversity was between 11% and 27%.

The distribution of certain species was interesting (Fig. 3). For example the dominance of *Anchomenus dorsalis* F. was 1.6-8.8% in organic orchards, 0.3-0.5 in IPM1, it was never found in IPM2 and C2, and only one specimen was found in C1 in 1992. This species has been described as typical of agro-ecosystems (Thiele 1977, Kromp 1989). However, it usually requires undisturbed habitat with permanent vegetation in which to overwinter (Thiele 1977, Kromp and Steinberg 1992, Paoletti and Sommaggio 1994). Similarly, there was a decline in *Amara aenea* Gyllh. numbers when comparing biological orchards to conventional ones. This may be due to the fact that biological orchards provide a high level of food for seed-feeding a Carabidae such as *A. aenea*.

Brachynus sclopeta F. was present in the organic orchards with a dominance of 0.7-14.2%; in integrated orchards dominance was 0.1-1.1. In conventional orchards only one specimen of this species was found in 1992. Several authors (e.g. Kromp 1989, Kromp and Steinberg 1992, Paoletti and Sommaggio 1994) observed that the *Brachynus* species usually occurs in the less disturbed habitats (e.g. undisturbed margins, hedgerows). Alderweireldt (1989) described *Pachignata deegeri* Sundevall as a species with an high preference for "hedge zones". It was found mainly in the organic orchards. Also *Forficula auricularia* L., with its polyphagous feeding habits (Glen 1975, Glen and Philips 1984), was present only in low numbers in integrated and conventional orchards.

The study of the earthworm population in all 64 orchards revealed a great reduction in biodiversity in the orchards with high and medium input (Fig. 4). Tillage had less effect on earthworm numbers than did chemical input. Tilled and untilled peach orchards had the same number of species of earthworms, while the tilled kiwi fruit orchards had one species more than untilled ones. Both tillage and chemical input, particularly the input of copper in the form of fungicidal sprays, reduced the total earthworm biomass (Fig. 5).

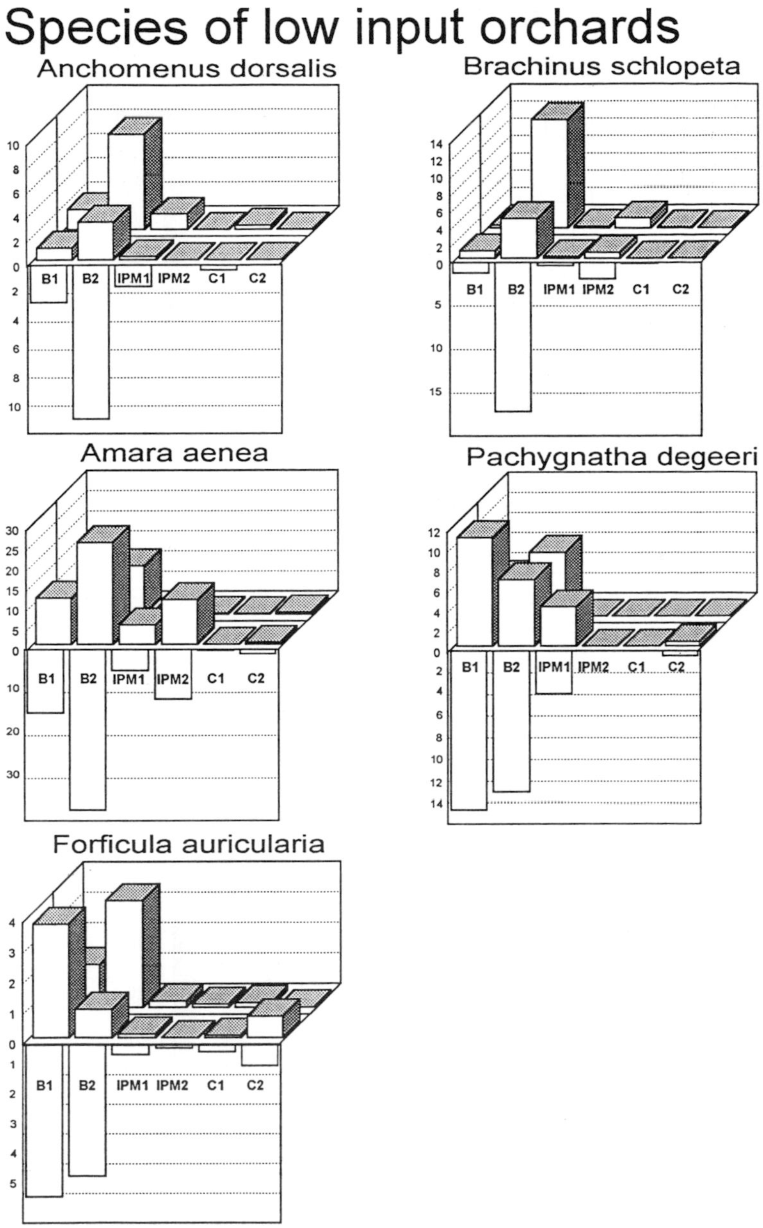
Species of low input orchards
Anchomenus dorsalis
Brachinus schlopeta
Amara aenea
Pachygnatha degeeri
Forficula auricularia
B1
B2
IPM1
IPM2
C1
C2

Fig. 3. Dominance of certain species. Values are the mean of ten sampling data in 1991 and eleven sampling data in 1992. Legend: B1 and B2: organic orchards; IPM1: integrated pest orchard with a spontaneous cover vegetation; IPM2: integrated pest management orchard with soil tillage; C1 and C2: conventional orchards.

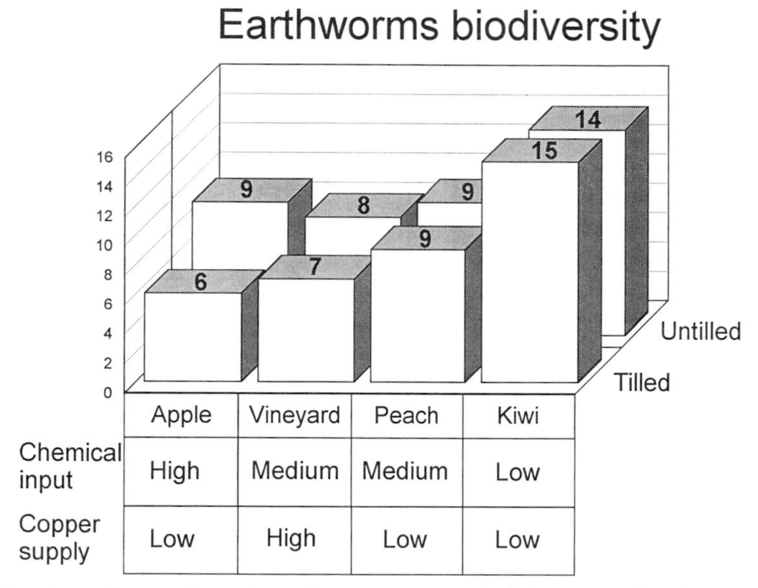

Fig. 4. Number of species of earthworms recorded in 64 orchards with different chemical input.

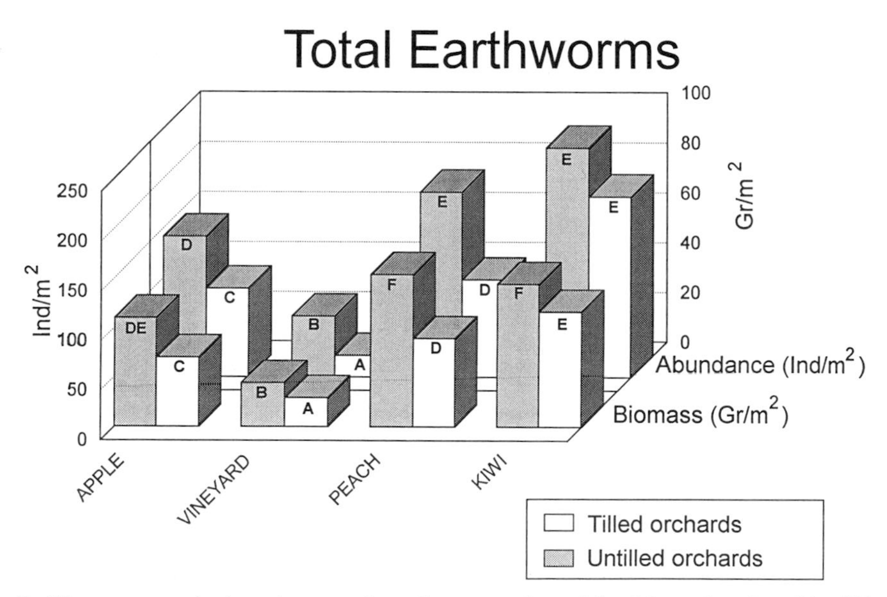

Fig. 5. Biomass and abundance of earthworms found in 64 orchards with different chemical input. Bars with different letters are significantly different (Kruskal-Wallis Test, $P < 0.05$).

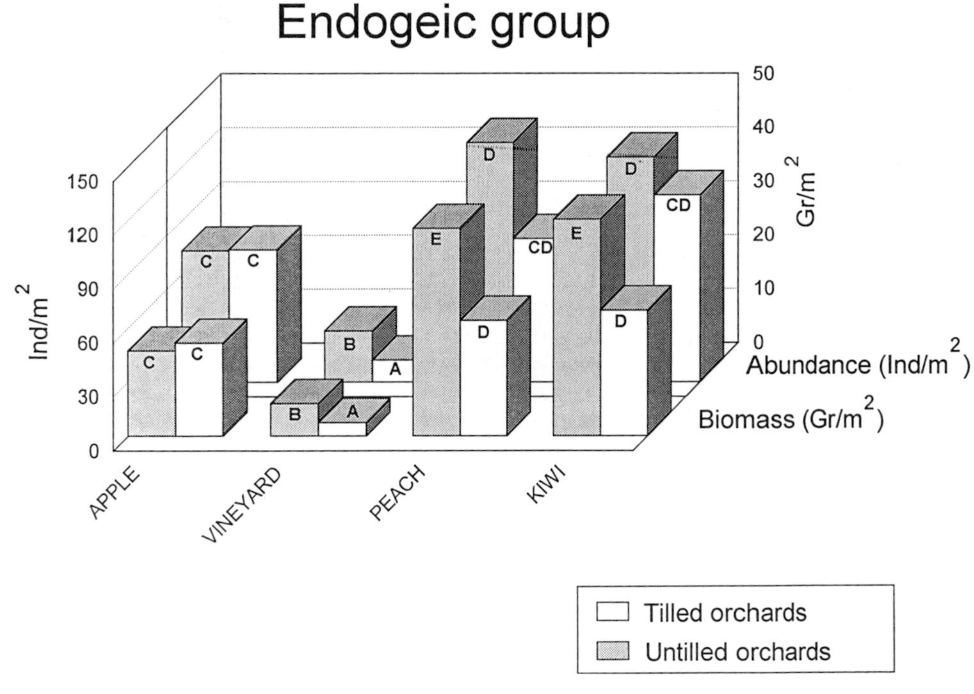

Fig. 6. Biomass and abundance of endogeic earthworm (Bouchè, 1977) recorded in 64 orchards with different chemical input. Bars with different letters are significantly different (Kruskal-Wallis Test, $P < 0.05$).

If we consider morphoecological categories of Bouchè (1977) an interesting behaviour can be observed. Epigeics and, in particular, *Lumbricus rubellus* Hoff and *Lumbricus castaneus* Sav., were severely reduced by soil tillage which destroyed litter layer where those species live. Similar results have been obtained by other researches, e.g. Kuhle (1983), Werner and Dindal (1989).

A 40-60% loss in endogeics earthworm abundance was observed in all orchards except in apple ones (Fig. 6). In particular *Aporrectodea caliginosa* Sav. showed a lower abundance and biomass in tilled orchards, while *Allolobophora chlorotica* Sav. were more abundant in tilled orchards, except for the vineyards. Particularly interesting was the response of epigeics and endogeics to the copper concentration in the soil. We failed to observe a reduction in epigeic abundance in the presence of high copper concentration in the soil (no difference has been observed between vineyards and other types of orchards). On the other hand, endogeics are severely reduced in the presence of high copper concentrations.

Endogeic earthworms and copper

y=a+b/lnx Adj r^2=0.46 r=0.68

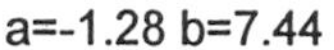

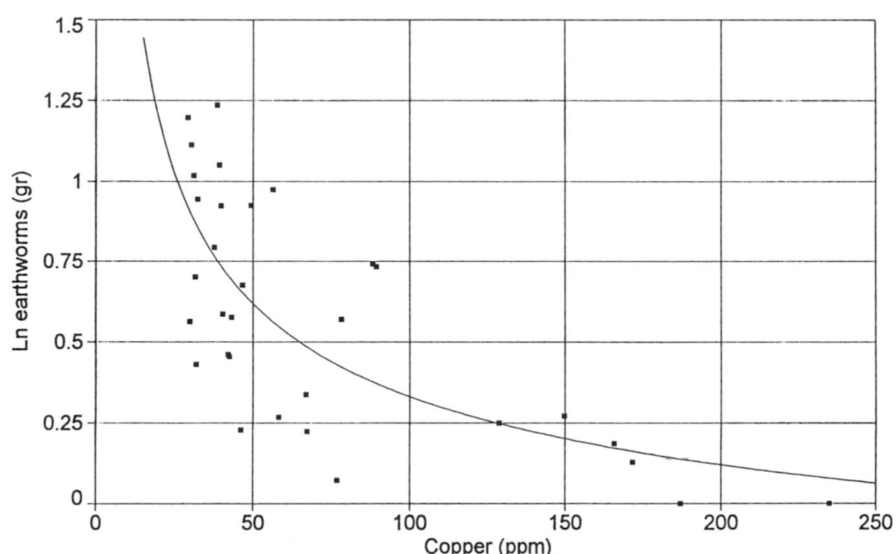

Fig.7. Regression between the copper content of the soil and the biomass (as logarithm) of endogeic earthworm in 32 orchards: Adj = coefficent of correlation adjusted for degrees of freedom. Each point is the mean value of 5 samplings for earthworm biomass and 4 samplings for copper concentration for each orchards.

They almost disappear when the copper increased above 150 ppm. A negative correlation has been found between copper concentration and endogeic biomass (Fig. 7). Copper content was responsible for 46% (R^2) of the endogeic distribution.

Discussion

"The agroecosystem differs from other 'wild' unmanaged climax ecosystems, in a similar physical environment, in being simpler, with less diversity of plant and animal species and with a less complex structure" (Tivy 1990, p.2). Agricultural practices typical of modern, conventional agriculture, e.g. monoculture, monosuccession and high chemical inputs, lead to a severe reduction in biodiversity (Paoletti 1988, Doran and Werner 1990, Altieri et al. 1987, Ryskowsky et al. 1991, Paoletti et al. 1992). Sustainable agriculture, with its reduced levels of soil disturbance, may produce an increase in biodiversity (e.g. Altieri et al. 1987, Paoletti et al. 1992).

Dritschilo and Erwin (1982) observed increased numbers of carabid species on organic farms. Kromp (1989) observed a 15-20% reduction of Carabidae species in conventional orchards. Booij and Noorlander (1992) and Booij (1990) stated that agricultural practice had less effect on Carabidae population than a change in crop type. El Titi and Ipach (1989) failed to record any increase in the numbers of earthworm species in integrated farms compared to conventional ones, but they did show a clear increase in earthworm biomass. Paoletti et al. (1995) observed an increase in both number of species and abundance of Lumbricidae in organic apple orchards when compared to conventional apple orchards; Carabidae and Isopoda abundance was clearly higher in the organic orchards, while the abundance of soil surface Araneae seemed to be correlated inversely with the amount of mechanical soil disturbance. In our research we observed a 30-39% increase in the number of species in biological orchards. Groups of invertebrates that seemed particularly sensitive to agricultural practices, including Isopoda, Lumbricidae and Araneae, were more numerous, as both species and number of individuals, in organic orchards. The presence in biological orchards of species typical of undisturbed habitats indicates how biological agricultural practices could increase the biodiversity in orchards. Those organisms that are sensitive to agriculture practice appear to be good bioindicators of the sustainability of agroecosystems.

Earthworms have been considered useful bioindicators because they are easy to collect and sensitive to agricultural practices (Kuhle 1983, Edwards and Bohlen 1992). Much research has recorded the high reduction of earthworms in the presence of high copper concentration (van Rhee 1967, 1969, 1975, Malecki et al. 1982, Ireland 1983, Cluzeau et al. 1987, Paoletti 1988, Paoletti et al. 1988). We found this response only for endogeics and only in a limited manner for epigeics. Several hypotheses can be advanced: such epigeics having high resistance, low exposure, higher recolonisation rates and the indirect effect of copper which can reduce soil decomposition, which is more important for endogeics than it is for epigeics. More research is needed to better detect such differences.

Our study confirms the importance of earthworms as bioindicators, allowing to a chemical stress (fungicide applications) to be distinguished from a mechanical one (tillage). Other invertebrates can, as well, give good indication to environmental stress (Isopoda, Carabidae, etc.), however their easier movement at the soil surface allow them to easily recolonize contaminated areas making additional complication to field preliminary approach.

Acknowledgements

We are indebted to Dr. C.J.H. Booij and M. Soratroi for improving our manuscript.

References

Alderweireldt, M. 1989. An ecological analysis of the spider fauna (Araneae) occurring in maize fields, Italian ryegrass fields and their edge zones, by means of different multivariate techniques. *Agric. Ecosystem Environ.* 27: 293-306.

Altieri, M.A., Anderson, M.K. & Merrick, L.C. 1987. Peasant agriculture and the conservation of crop and wild plant resources. *Conserv. Biol.* 1: 49-58.

Booij, K. 1994. Diversity patterns in carabid assemblages in relation to crops and farming systems. In: Desender, K., Dufrene, M., Loreau, M., Luff, M.L. & Maelfait, J-P. (eds.) *Carabid beetles: Ecology and Evolution.* Kluwer Academic Publishers, Dordrecht/Boston/London, 425-31.

Booij, C.J.H. & Noorlander, J. 1992. Framing systems and insect predators. *Agric. Ecosystems Environ.* 40: 125-35.

Bouchè, M.B. 1977. Strategies lombriciennes. In: Lohm, U. & Persson, T. (eds.) *Soil organisms as components of ecosystems*, Biological Bolletin, Stockholm, 25: 122-32.

Brandmayer, P. & Brunello-Zanitti, C. 1982. Le comunità di Coleotteri Carabidi in alcuni querco-carpineti della Bassa Pianura del Friuli. *CNR*, AQ/1/181-186, 69-124.

Cluzeau, D., Lebouvier, M., Trahen, P., Bouchè, M.B., Badour, C. & Perraud, A. 1987. Relations between earthworms and agricultural practices in the vineyards of Champagne. Preliminary results. In: Bonvicini Pagliai, A.M. & Omodeo, P. (eds.) *On earthworms*, Mucchi Modena, 465-84.

Doran, J.D. & Werner, M.R. 1990. Management and soil biota. In: Francis, C.A., Flora, C.B. & King, L.D. (eds.) *Sustainable Agriculture in Temperate Zones*, Wiley, N.Y. 205-30.

Dritschilo, W. & Erwin, T.L. 1982. Responses in abundance and diversity of cornfield carabid communities to differences in farm practices. *Ecology* 63: 900-4.

Edwards, C.A. & Bohlen, P.J. (1992). The effects of toxic chemicals on earthworms. *Reviews of Environmental Contamination and Toxicology* 125: 23-99.

El Titi, A. & Ipach, U. 1989. Soil fauna in sustainable agriculture: results of an integrated farming system al Lautenbach, F.R.G. *Agric. Ecosystems Environ.* 27: 561-72.

Ehrlich, P.R. & Wilson, E.O. 1991. Biodiversity studies: science and policy. *Science* 253: 758-62.

Erwin, T.L. 1988. The tropical forest canopy – the heart of biotic diversity. In: Wilson, E.O. (ed.) *Biodiversity.* National Academic Press, Washington, DC, 3-18.

Glen, D.M. 1975. The effects of predators on the eggs of codling moth *Cydia pomonella*, in cider-apple orchard in south-west England. *Ann. Appl. Biol.* 80: 115-35.

Glen, D.M. & Philips, M.L. 1984. Integrating control of Erwings. *Grower*, Bristol, 7: 35-37.

Ireland, M.P. 1983. Heavy metal uptake and tissue distribution in earthworms. In: Satchell, J.E. (ed.) *Earthworm ecology from Darwin to vermiculture*, Chapmann and Hall, London, 247-65.

Kromp, B. 1989. Carabid beetle communities (Carabidae, Coleoptera) in biologically and conventionally farmed agroecosystems. *Agric. Ecosystems Environ.* 27: 241-51.

Kromp, B. & Steinberger, K.H. 1992. Grassy field margins and arthropod diversity: a case study on ground beetles and spiders in eastern Austria (Coleoptera: Carabidae; Arachnida: Aranei, Opiliones). *Agric. Ecosystems Environ.* 40: 71-93.

Kuhle, J.C. 1983. Adaption of earthworm populations to different soil treatment in apple orchards. In: Lebrun, Ph., Andrè, H.M., De Medts, A., Gregoire-Wibo, C. & Wauthy, G. (eds.) *Proceedings of the VIII International Colloquium of Soil Zoology*, Louvain-la-Neuve (Belgium). August 30 – September 2, 1982, 487-501 Dieu-Brichart, Ottignies-Louvain-la-Neuve.

May, R. 1992. How many species inhabit the earth? *Sci. Am.* 267: 42-48.

Malecki, S.H.P., Neuhauser, E.F. & Loher, R.C. 1982. The effect of metals on the growth and reproduction of *Eisenia foetida* (Oligochaeta, Lumbricidae). *Pedobiologia* 24: 129-37.

Paoletti, M.G. 1988. Soil Invertebrates in cultivated and uncultivated soils in North Eastern Italy. *Redia* 71: 501-63.

Paoletti, M.G., Favretto, M.R., Marchiorato, A., Bressan, M., e Babetto, M. 1993. Biodiversita in pescheti forlivesi. In: Paoletti, M.G., Favretto, M.R., Nasolini, P., Scaravelli, D. e Zecchi, M. (eds.) *Biodiversita negli agroecosistemi*, Osservatorio Agroambientale, Centrale Ortofrutticola, Forli, 20-56.

Paoletti, M.G., Pimentel, D., Stinner, B.R. & Stinner, D. 1992. Agroecosytem biodiversity: matching production and conservation biology. *Agriculture, Ecosystems and Environments* 40: 3-23.

Paoletti, M.G., Iovane, E. & Cortese, M. 1988. Pedofauna bioindicators and heavy metals in five agroecosystems in north-east Italy. *Revue Ecologie et Biologie du Soil* 25: 33-58.

Paoletti, M.G., Schweigl, U. & Favretto, M.R. 1995. Soil macroinvertebrates, heavy metals and organochlorines in low and high input apple orchards and a coppiced woodland. *Pedobiologia*, in press.

Paoletti, M.G. & Sommaggio, M.G. 1994. Preliminary data of margin effects on soil surface Carabids in Italy (1993 report). *Field Margins Newsletter*, Long Asthon Research Station, Number 3.

Raven, P. 1992. The Nature and Value of Biodiversity. In: *WRI, IUCN, UNEP, FAO, UNESCO Global Biodiversity Strategy*, 1-5.

Rhee, J.A. van. 1967. Development of earthworm populations in orchards soils. In:

Graff, O. & Satchell, J.E. (eds.) *Progress in soil biology*, North Holland Publishing Company, Amsterdam, 360-71.

Rhee, J.A. van. 1969. Effects of biocides and their residues on earthworms. *Med. Ryksfac. Landbouww.* Gent, 34: 682-89.

Rhee, J.A. van. 1975. Copper contamination effects on earthworms by disposal of pig waste in pastures. In: Vanek, J. (ed.) *Progress in Soil Zoology*, Proocedings 5th International Colloquium of Soil Zoology, Prague, 1973.

Ryskowski, L., Karg, J., Margarit, G., Paoletti, M.G. & Zlotin, R. 1993. Above ground insects biomass in agricultural landscapes of Europe. In: Bunce, R.G.H., Ryskowski, L. & Paoletti, M.G. (eds.) *Landscape Ecology and Agroecosystem Trends*, Lewis Publishers, Boca Raton/Ann Arbor/London/Tokyo, 71-82.

S.I.S.S. 1985. *Metodi Normalizzati di Analisi del Suolo.* Edagricole.

Stork, N.E. 1988. Insect diversity: facts, fiction and speculation. *Biol. Journal Linnean Society* 35: 321-37.

Thiele, H.U. 1977. Carabid beetles in their environments. *Zoophysiology and Ecology* Vol.10 Springer Verlag, Berlin/Heidelberg/New York.

Werner, M.R. & Dindal, D.L. 1989. Earthworm community dynamics in conventional and low-input agroecosystems. *Rev. Ecol. Biol. Soil* 26: 427-37.

Wheeler, Q. 1990. Insect diversity and cladistic constrains. *Ann. Entomol. Soc. Am.* 83: 1031-47.

Wilson, E.O. 1988. The current state of biological diversity. In: Wilson, E.O. (ed.) *Biodiversity*, National Academic Press, Washington, DC, 3-18.

Ground photoeclector evaluation of the numbers of carabid beetles and spiders found in and around cereal fields treated with either inorganic or compost fertilizers

J. Idinger[1], *B. Kromp*[1]
& *K.-H. Steinberger*[2]

[1] L. Boltzmann-Institute for Biological Agriculture and Applied Ecology
Rinnböckstrasse 15, A-1110 Vienna, Austria
[2] Institute of Zoology, University of Innsbruck
Technikerstrasse 25, A-6020 Innsbruck, Austria

Abstract
In 1991 and 1992, carabid beetles and spiders that emerged from the soil were sampled by ground photoeclectors in different parts of inorganically grown rye crop. Samples were taken in an unfertilized plot, a compost-fertilized plot, alongside a hedgerow and at the edge of a forest. Samples were also taken from a nearby field treated with inorganic fertilizer.

Similar numbers of beetle and spiders were recorded from each plot. Only in the field treated with inorganic fertilizer less beetles and spiders were caught. In carabids, but not in spiders, diversity decreased with fertilization. In the adjacent habitats, more spider and carabids species were found than in the field plots. *Trechus quadristriatus* was the most abundant carabid species in the field treated with inorganic fertilizer. For *Bembidion lampros, Asaphidion flavipes* and *Platynus dorsalis*, migrations between field margins and fields were indicated by the seasonal changes in their distribution. In total 34 carabid species and 78 spider species were caught.

Differences in carabid and spider faunas between plots are considered to arise from the indirect effects of fertilizers on soil- and crop-related properties, as well as from the different numbers of prey available to the various predators.

Key words: carabidae, spiders, organic farming, fertilizers, field margins, rye, photoeclector

---

Arthropod natural enemies in arable land · II *Survival, reproduction and enhancement*
C.J.H. Booij & L.J.M.F. den Nijs (eds.). *Acta Jutlandica* vol. 71:2 1996, pp. 255-267
 ISBN 87 7288 672 2

Introduction

Non-specialist ground predators like carabids and spiders are considered to play an important role in pest control in field crops (see reviews of Riechert & Lockley 1984, Luff 1989). Under different farming systems, although predatory arthropod numbers appear to be increased by organic cultivation (see e.g. Ingrisch et al. 1989), the influence of the individual components involved is poorly understood.

With respect to organic fertilization, some data exist on the effect of manure in increasing the numbers of carabid beetles (Pietraszko & De Clercq 1982, Purvis & Curry 1984, Hance & Gregoire-Wibo 1987).

In contrast, data is not available on how compost fertilizers affect the numbers of predatory arthropods, particularly composts produced from separately collected organic components of household rubbish. The city of Vienna currently produces about 30,000 tons per year of such compost. The effect of this compost in organic systems of crop production has been tested experimentally by the L. Boltzmann-Institute (see Amlinger 1993). This work has been accompanied by parallel studies on the arthropod faunas of such fields and their adjacent habitats (e.g. Kromp & Steinberger 1992, Kromp & Nitzlader, in press). Idinger (1994) used ground photoeclectors to compare how treating plots with either inorganic nitrogen or compost fertilizers affected the arthropod fauna. She also included carabid beetles and spiders, which together made up about 6% of the total arthropods collected. The spiders were identified to the species level by Steinberger (1994 unpubl.). Results of the numbers of both carabid beetles and spiders collected from sites within and alongside a crop of rye grown organically are presented in this paper.

Material and Methods

The study was conducted on a conversion farm (48°10'N and 16°30'E; 152 m above sea level; 9.6°c, 510 mm; greyish alluvial soil) at Obere Lobau, Vienna, that formed part of a riverside nature reserve.

Sampling was done using ground photoeclectors. These consisted of a 0.25 m^2, square metallic frame, in which the removable upper part was covered by green raincoat material. A single pitfall trap, that was placed into the soil beneath the frame, contained 2% formaldehyde to preserve the trapped arthropods. The upper trap had a transparent cover and contained 1% formaldehyde. Six samples were taken between May and early November in

1991 and five between May and August in 1992. Each sample consisted of the number of arthropods collected over a 14-day period.
Samples were taken from the following areas:
– N: a field plot (185 x 10 m) that had not been fertilized since autumn 1989
– C: a field plot (185 x 10 m) that had been fertilized with 80 t/ha of compost in the autumns of both 1989 and 1991. Both plots were situated in the central part of a 4 ha organic field (crops: winter-wheat in 1990, winter-rye in 1991 and 1992) and were plots from a trial for testing the effects of applying varying amounts of compost on crop growth (Amlinger & Walter 1993)
– I: a field (7.6 ha) situated nearby (crops: potatoes/green beans in 1990, winter-wheat in 1991, winter-rye in 1992) in which herbicide was sprayed and inorganic fertilizer applied in 1990/91 (autumn 1990: 30 N, 75 P, 120 K spring 1991: 112 N, 104 Ca; amounts in kg/ha). This field was being converted to organic cultivation from 1991/92 onwards, and therefore herbicides and inorganic fertilizers were not applied in the 2nd year of the investigation. All sites were sprayed with wettable sulphur in both years and similar methods were used for cultivating the soil on all sites
– H: the grassy margin of a hedge
– F: the grassy margin of a forest. Both H & F were adjacent to the field containing the plots N and C.

In the middle of each of the three field plots, five ground photoeclectors were placed in a line and spaced approximately 20 m apart. Three traps were placed also in a line, at intervals of 30-40 m, both alongside the selected hedgerow and the edge of the forest.

Extensive information on the soil characteristics of the sites is given by Idinger (1994) (e.g. soil moisture, particle size distribution, pH, Norg, Nmin, humus, C/N ratio, macronutrients, cation exchange capacity, macronutrients, heavy metals), crop development (crop coverage, stalk lengths), weed vegetation (weed coverage, weed species), and possible prey arthropods (e.g. collembolans, dipterans) for carabids and spiders associated with each field plot used in this series of experiments.

Results

Carabid beetles

In total, 1258 carabid beetles, representing 34 species, were collected during 1991 and 1992. Table 1 includes data on the numbers of beetles trapped, the total species caught and Shannon's diversity H' (calculated according to Magurran, 1988).

No clear differences could be detected in the numbers of beetles caught in the field sites, where values ranged from a mean of 4.2 per trap in the compost plot to 4.6 in the field treated with inorganic fertilizer. Of the total of 23 species of carabid beetles caught in the field plots, 18 spp. were collected from both the plot fertilized with compost and the non-fertilized plot, whereas only 12 spp. were caught in the field in which inorganic

Table 1. Total numbers of carabid beetles collected in 1991 and 1992 at Obere Lobau, Vienna, from three differently fertilized field sites and two adjacent field margins. In 1991 six and in 1992 five photoeclector samples of 14 day-periods were taken throughout the season.
I = inorganic-fertilized, C = compost-fertilized, N = not-fertilized
H = grassy hedge-margin, F = grassy forest edge.

Research sites	I	C	N	H	F	Total
# photoeclectors	5	5	5	3	3	21
species						
Bembidion lampros	62	103	82	60	139	446
Trechus quadristriatus	134	60	46	1	10	251
Syntomus obscuroguttatus	5	10	19	69	76	179
Asaphidion flavipes	12	15	14	7	24	72
Platynus dorsalis	3	8	21	16	19	67
Amara plebeja	2	7	4	12	42	67
Stomis pumicatus		2	22	4		28
Demetrias atricapillus	1	4	8	3	3	19
Brachinus explodens			6	7	6	19
Poecilus cupreus	6	6	2		1	15
Panageus bipustulatus	2			3	9	14
Drypta dentata		2	8	3		13
other species	3	11	12	11	31	68
total individuals	230	228	244	196	360	1258
mean # /0.25 m2/14 days	4.6	4.2	4.4	5.9	10.9	
total # species	12	18	18	16	24	34
Shannon's H (ln)	1.23	1.76	2.14	1.89	2.01	

Table 2. Total numbers of spiders collected in 1991 and 1992 at Obere Lobau, Vienna, from three differently fertilized field sites and two adjacent field margins. In 1991 six and in 1992 five photoeclector samples of 14 day-periods were taken throughout the season.
I = inorganic-fertilized, C = compost-fertilized, N = not-fertilized
H = grassy hedge-margin, F = grassy forest edge.

Research sites	I	C	N	H	F	Total
# photoeclectors	5	5	5	3	3	21
species						
Erigona atra	71	103	87			261
Meioneta rurestris	46	87	91	3	1	228
Araeoncus humilis	69	49	51	1		170
Erigona dentipalpis	23	29	31	1		84
Bathyphantes gracilis	22	42	14	1		79
Neottiura bimaculata	20	18	13	3	2	56
Oedothorax apicatus	11	13	30			54
Porrhoma microphthalmum	26	17	7	1	1	52
Pardoda lugubris		6	5	6	1	18
Syedra gracilis				15	1	16
Ballus depressus		1		9	1	11
Zelotes pedestris				4	7	11
Walckenaera vigilax	5		5			10
Leptyphantes tenuis		1	3	4	2	10
Oxyptila praticola				4	6	10
Haplodrassus silvestris				7	3	10
Tetragnatha pinicola	4	3	1	1		9
Enoplognatha schaufussi	5	1	2			8
Pardosa agrestis		4	3			7
Dictyna incinata				3	4	7
Leptyphantes pallidus				3	4	7
other species	16	18	19	39	38	130
total individuals	318	392	362	105	71	1248
mean # /0.25 m^2/14 days	6.4	7.1	6.6	3.2	2.2	
total # species	21	30	30	37	37	78
Shannon's H' (ln)	2.26	2.23	2.31	3.28	3.39	

fertilizer was applied. Similarly, the diversity values estimated for the carabid beetles decreased from the non-fertilized plot (H': 2.14) to the compost-fertilized plot (H': 1.76) and finally to the field treated with inorganic fertilizer (H': 1.23).

More carabid beetles were caught in the field margins than in the actual fields. Most beetles (11/trap) were caught at the grassy forest edge. The highest number (24 spp.) of species was also recorded from the edge of the forest. The relative numbers of the dominant carabid beetles collected in the field plots, declined gradually from the field treated with inorganic fertilizer through the compost-treatment and the non-treated plot towards the margins of the hedgerow (Fig. 1). *Trechus quadristriatus*, is clearly a field-species that aggregates in fields treated with inorganic fertilizer. In contrast, other species, such as *Asaphidion flavipes*, *Platynus dorsalis*, *Amara plebeja* and *Syntomus obscuroguttatus*, occurred in low numbers in the field treated with inorganic fertilizer, but increased gradually in numbers towards the more established habitats.

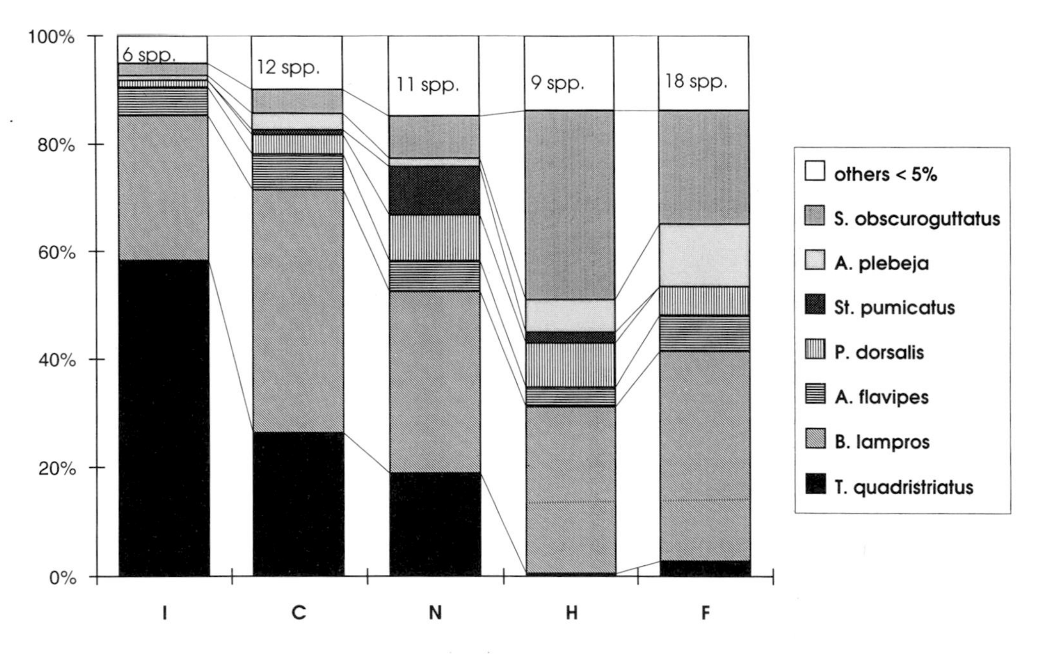

Fig. 1. Relative composition of carabid beetle populations in three field sites (I, C, N) and two adjacent habitats (H, F), sampled by ground photoeclectors in 1991 and 1992 at a conversion farm in Obere Lobau, Vienna. (Abbreviations: I = inorganic-fertilized, C = compost-fertilized, N = not-fertilized; H = grassy margin of a hedgerow, F = grassy forest edge)

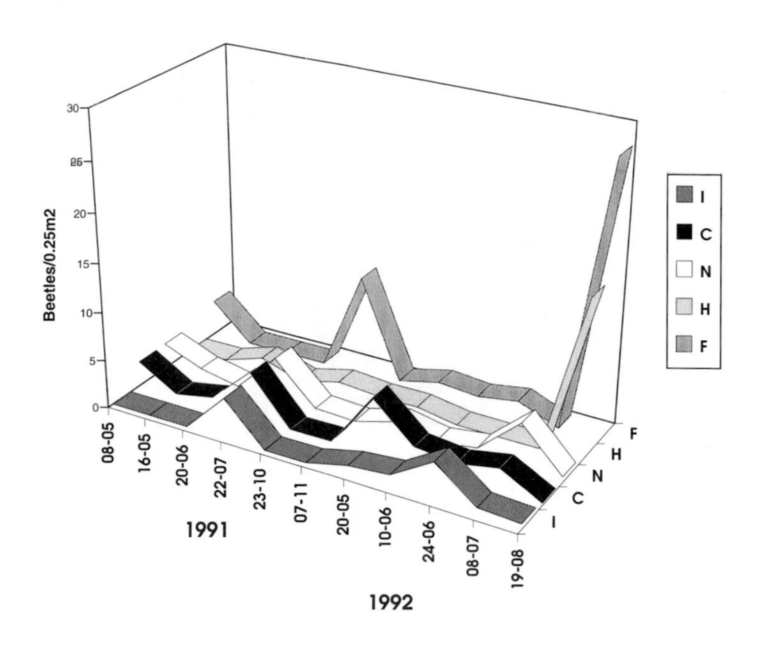

Fig. 2. Seasonal changes in numbers of *Bembidion lampros* (Obere Lobau, Vienna; 1991/92). (For abbreviations of sampling sites, see Fig. 1)

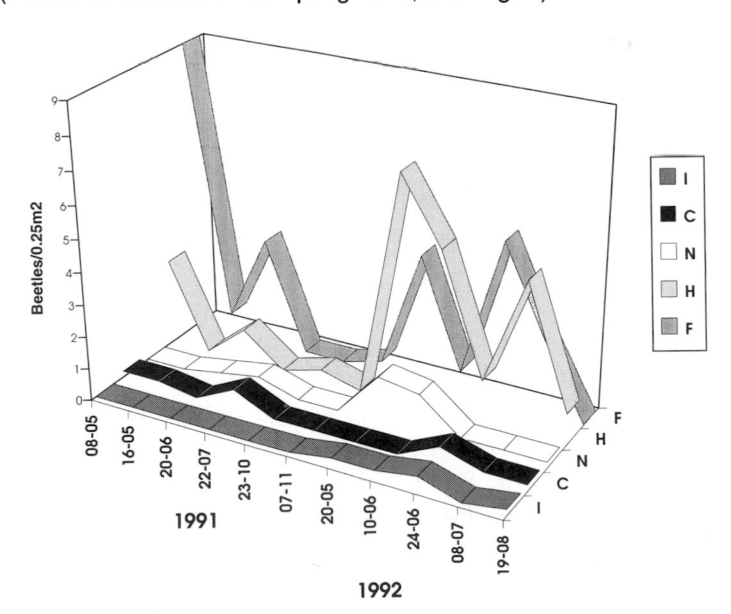

Fig. 3. Seasonal changes in numbers of *Syntomus obscuroguttatus* (Obere Lobau, Vienna; 1991/92). (For abbreviations of sampling sites, see Fig. 1)

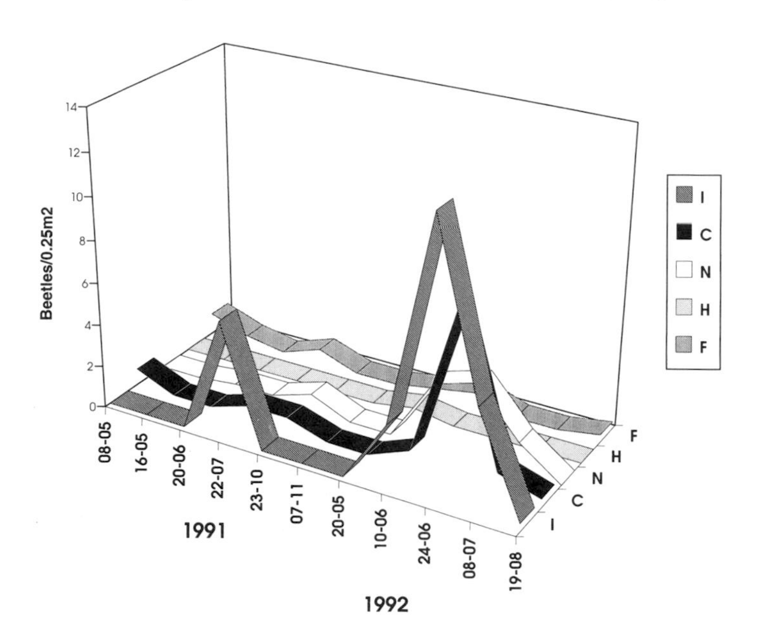

Fig. 4. Seasonal changes in numbers of *Trechus quadristriatus* (Obere Lobau, Vienna; 1991/92). (For abbreviations of sampling sites, see Fig. 1)

Although high numbers of *Bembidion lampros* were collected from all sites, most were collected at the forest edge (Fig. 2). The time delay in capturing beetles along the field margins and the plots provides a measure of beetle dispersal (see also Kromp & Steinberger 1992). About 50% more *B. lampros* were caught in the plot treated with compost in 1992 than in 1991 (data in Idinger 1994). In contrast, numbers were similar in both years in the non-fertilized plot. Migrations between habitats was also indicated by the data collected for *P. dorsalis*.

S. obscuroguttatus (Fig. 3) appears to be a true "edge species" (Kromp & Nitzlader, in press), being caught only in low numbers in the field sites. In contrast, *T. quadristriatus* is considered a typical "field species" as it occurred in high numbers in the field plots and only low numbers in the field margins (Fig. 4). Newly-emerged adults were quite common in the head traps, as were tenerals of *S. obscuroguttatus* and *Demetrias atricapillus*, indicating that these three species disperse by flying.

Spiders

The 1248 adults caught represented 78 spp. from 14 families (Table 2). In addition, 3403 juveniles were caught, among which there were 5 spp. and 7 families that were not detected as adults (Steinberger, 1994, unpubl.). The juvenile data were not evaluated further, as this type of trapping is not considered appropriate for estimating the density of juveniles. Therefore, the following results refer to adult spiders only.

Like the carabid beetles, the numbers of spiders caught showed no clear differences between the field plots, and ranged from 6.4 in the plot treated with inorganic fertilizer to 7.2 individuals/0.25 m^2/14 days in the plot treated with compost. The numbers of species caught were also lower in the field treated with inorganic fertilizer (21 spp.), than in the plots treated with compost or not treated (30 spp. each). Unlike the carabid beetles, the application of fertilizer did not appear to influence spider diversity, as it was similar in all three plots (see bottom line of Table 2).

The dominance structure of the spider coenoses was quite different to that of the carabid beetles (Fig. 5). In all three field plots, seven linyphiid species (*Erigone atra, Meioneta rurestris, Araeoncus humilis, Erigone dentipalpis, Bathyphantes gracilis, Oedothorax apicatus, Porrhomma microph-*

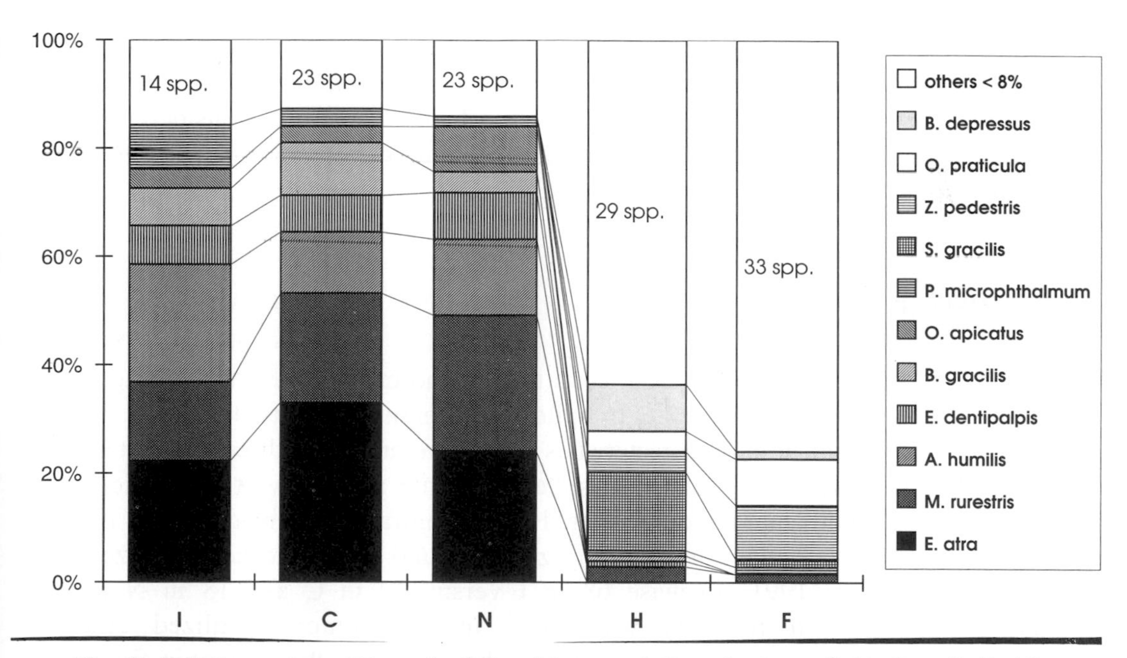

Fig. 5. Relative composition of adult spider populations in three field sites (I, C, N) and two adjacent habitats (H, F), sampled by ground photoeclectors in 1991 and 1992 at a conversion farm in Obere Lobau, Vienna. (For abbreviations of sampling sites, see Fig. 1)

thalmum) made up about 85% of total spiders caught, each species having a similar dominance ranking across all sites. The number of additional species, however, was clearly higher in the plot treated with compost and the untreated plot than in the field treated with inorganic fertilizer. Among these spiders, the Lycosidae were represented by four *Pardosa* species that were trapped only in the compost and non-treated plots (Steinberger 1994, unpubl.).

Comparison of the numbers of spiders caught in the compost and non-fertilized plots in 1991 and 1992 (Steinberger 1994, unpublished data), revealed a 26% increase in numbers in the compost plot in 1992. In both years, spider numbers were similar in the non-fertilized plots.

Compared to carabid beetles, the spider populations within the fields and along the field margins differed considerable at Obere Lobau. Although fewer individuals were caught in the margins (hedge: 3.2, forest edge: 2.2), the total species caught (hedge and forest edge: 37 spp. each) and Shannon's diversity H' (hedge: 3.3, forest edge: 3.4) were much higher. The typical agrotopic species that were abundant in the fields, were caught in very low numbers alongside the hedgerow and forest edge. In these locations, many other species were caught, often only in low numbers, and included typical ecotone (Steinberger, unpubl., e.g. *Dictyna uncinata, Oxyptila praticola, Pardosa lugubris*) and woodland species (e.g. *Haplodrassus silvestris, Leptyphantes pallidus*). The grassy margin of the hedge, is considered a highly suitable microhabitat for xero-thermic spiders, e.g. *Syedra gracilis*, which is a rare species (Steinberger, unpubl.).

Discussion

To help interpret the faunistic data, certain soil- and crop-related data (Idinger 1994) will be mentioned briefly. In spite of different amounts and types of fertilizer, the chemical soil characteristics were similar for all three field sites, except for the nitrate level, which was clearly higher after harvest in the field treated initially with inorganic fertilizer. Here the nitrate level reached 100 kg NO3/ha versus 51 in the compost-fertilized plot and 31 in the non-fertilized plot in November 1991 (likewise 68 in I versus 29 in C and 18 in N in autumn 1992). Soil moisture was equal in the compost-fertilized and non-fertilized plots. The values were somewhat lower in the inorganic field, possibly due to higher water-consumption by the more actively-growing crop. For rye in 1992, the highest crop coverage and stalk lengths were found in the field treated with inorganic fertilizer, being intermediate in the compost-treated plot and lowest in the non-fertilized plot. In contrast, weeds were more prominent in the non-fertilized plot, followed by the compost plot and the field treated with inorganic fertilizer. Weed species were clearly reduced in the latter field in 1991 due to the herbicide applied (7 spp. versus 14 in C and U), but soon recovered to their former value in 1992 (19 spp. versus 20 in the other fields) when herbicide was not applied.

Some arthropod groups showed significant differences in numbers between the field sites. For example, in 1992, isotomid collembolans, chironomid and sciarid midges as well as drosophilid flies were most abundant in the plot treated with compost, whereas most aphids were recorded in the field treated with inorganic fertilizer (Idinger 1994).

Overall, carabid beetles seem to be more sensitive than spiders to changes in the amount and the type of fertilizer applied to field crops. Differences in numbers of both carabid beetles and spiders, are considered to result from fertilizer-induced, complex indirect changes in the soil biota and in the structure and microclimate of the vegetation of both the crop plants and weeds.

At present, no direct relationship has been derived between nitrate levels and the numbers of carabid beetles and spiders found in crop fields. The only publications on the effects of nitrogen fertilizers on predatory arthropods are those by Honczarenko (1975) and Kajak (1981), who applied high doses of nitrogen to meadows. After eight years of application, Kajak (1981) found that large, mobile lycosids were replaced by smaller, sedentary linyphiid spiders. Lycosid spiders are also more abundant in organic fields than in fields cultivated conventionally (Glück & Ingrisch 1990, Steinberger & Kromp 1993). In this study, lycosids were not present in the samples collected from the field treated with inorganic fertilizer.

The higher nitrogen supply in the field treated with the inorganic fertilizer enabled better crop development; and the denser crop stand created a shadier and more humid microclimate. According to Mitchell (1963), the latter is preferred by *T. quadristriatus*, a species considered to be enhanced by intensive conventional cultivation (see Kromp 1990).

In the compost-fertilized and non-fertilized plots, the lower level of nitrogen resulted in a warmer and drier microclimate, which, together with greater heterogeneity of the crop stand and a higher weed diversity, possibly provided additional niches for less-abundant species of both spiders and carabid beetles. Such conditions might also have enhanced the subdominant species of carabid beetles mentioned earlier. In conclusion, the increasing similarity of the carabid beetle population within the crop and in the field margins, described earlier by Kromp & Steinberger (1992) for an organic wheat field and a grassy field margin, is considered to be related to the reduced amounts of fertilizer available.

The carabid coenosis of the compost-fertilized plot showed intermediate dominance of *T. quadristriatus* and subdominant species compared with inorganically-fertilized and non-fertilized plots (see Fig. 1). In contrast, the spider population had almost the same relative composition in all three field plots (see Fig. 5). The increase in the total numbers of spiders and *B. lampros* caught in the compost plot in 1992, was possibly related to the application of compost in autumn 1991, increasing the numbers of arthropod prey items. According to Glück & Ingrisch (1990), collembolans and dipterans, together with aphids and hemipterans, are the main food sources of field erigonid and lycosid spiders. *B. lampros* is known to feed, amongst other arthropods, on eggs of brachyceran flies (see Luff 1989). Purvis & Curry (1984) found increased *B. lampros* abundances in manured versus non-manured plots. Likewise, Hance & Gregoire-Wibo (1987) reported *B. lampros* to be dominant in sugar beet crops only after organic material had been applied to such crops.

In conclusion, at Obere Lobau, differences in the numbers of carabid beetles and spiders collected were more pronounced between the plots treated with inorganic fertilizer and compost than between the compost plots and the plots that were not treated with fertilizer. Evidence that the application of inorganic fertilizers affects the predatory arthropods is probably biased by the impacts of the other cultivation measures that were used, such as the different crop rotations and the fact that herbicide was applied only in the field that was treated with inorganic fertilizer.

Although the addition of compost had little effect on the numbers of carabid beetles and spiders caught in the present study during the 2nd and 3rd year of compost application, the effects are expected to increase in subsequent years. In general, the beneficial effect of compost on the soil biota is considered to occur rather slowly. A future study will evaluate arthropod numbers in a replicated plot trial that contains a range of both nitrogen treatments and compost fertilizers.

Acknowledgements

This work was funded by a grant from the Austrian Ministry of Science and Research.

References

Amlinger, F. 1993. *Biotonne Wien – Theorie und Praxis.* MA 48 – Stadtreinigung und Fuhrpark, Wien, 385 pp.

Glück, E. & Ingrisch S. 1990. The effect of bio-dynamic and conventional agriculture management on Erigoninae and Lycosidae spiders. *J. Appl. Ent.* 110: 163-48.

Hance, Th. & Gregoire-Wibo, C. 1987. Effect of agricultural practices on carabid populations. *Acta Phytopath. Entom. Hung.* 22: 147-60.

Honczarenko, J. 1975. An influence of high-dose nitrogen fertilizer on the Entomofauna of meadow soil. *Pedobiologia* 16: 58-62 (in Russian, with English summary).

Idinger, J. 1994. Untersuchungen zur Schlüpftrichterfauna der in Umstellung auf biologischen Landbau befindlichen Getreidefelder in der Oberen Lobau/Wien unter spezieller Berücksichtigung von Kompostdüngern und Ausgleichsbiotopen. Ph.D. Thesis, University of Vienna, 234 pp.

Ingrisch, S., Glück, E. & Wasner, U. 1989. Zur Wirkung des biologisch-dynamischen und konventionellen Landbaues auf die oberirdische Fauna des Ackers. *Verh. Ges. Ökol.* 18: 835-41.

Kajak, A. 1981. Analysis of the effect of mineral fertilization on the meadow spider community. *Ekologia Polska* 29: 313-26.

Kromp, B. 1990. Carabid beetles (Coleoptera, Carabidae) as bioindicators in biological and conventional farming in Austrian potato fields. *Biol. Fertil. Soils* 9: 182-87.

Kromp, B. & Steinberger, K.H. 1992. Grassy field margins and arthropod diversity: a case study on ground beetles and spiders in eastern Austria (Coleoptera: Carabidae; Arachnida: Aranei, Opiliones). *Agric. Ecosystems Environ.* 40: 71-93.

Luff, M.L. 1989. Biology of polyphagous ground beetles in agriculture. *Agric. Zool. Rev.* 2: 237-78.

Magurran, A.E. 1988. Ecological diversity and its measurement. Croom Helm, London/Sydney, 179 pp.

Mitchell, B. 1963. Ecology of two carabid beetles, *Bembidion lampros* (Herbst) and *Trechus quadristriatus* (Schrank). II. Studies on populations of adults in the field, with special reference to the technique of pitfall trapping. *J. Anim. Ecol.* 32: 377-92.

Pietraszko, R. & De Clercq, R. 1982. Influence of organic matter on epigeic arthropods. *Med. Fac. Landbouww. Rijksuniv. Gent* 47/2: 721-28.

Purvis, G. & Curry, J.P. 1984. The influence of weeds and farmyard manure on the activity of Carabidae and other ground-dwelling arthropods in a sugar beet crop. *J. Appl. Ecol.* 21: 271-83.

Riechert, S.E. & Lockley, T. 1984. Spiders as biological control agents. *Ann. Rev. Entomol.* 29: 299-320.

Steinberger, K.H. & Kromp, B. 1993. Barberfallenfänge von Spinnen in biologisch und konventionell bewirtschafteten Kartoffelfeldern und einer Feldhecke bei St. Veit (Kärnten, Österreich) (Arachnida: Aranei). *Carinthia* II, 183./103.: 657-66.

Agricultural practices which enhance numbers of beneficial arthropods

H.M. Poehling

Universität Hannover, Institut für Pflanzenkrankheiten und Pflanzenschutz, Herrenhäuser Str. 2, D-30419 Hannover, Germany

Introduction

This paper is based upon discussions held during an EC-Workshop entitled "Survival, reproduction and enhancement of beneficial predators and parasitoids in agroecosystems" which took place in Wageningen, The Netherlands in December 1994.

This paper should give only a short summary of the main discussion items and contributions. It is not intended to give a comprehensive literature review. The overall structure of the discussion, particularly the evaluated subjects, tools and objectives are illustrated in Fig. 1. The complex of papers was divided into some small chapters which are dealt with below under different headlines. However, it has to be mentioned that the topics are closely interlinked and some aspects are partially repeated.

System approaches – Integrated farming

The evaluation of complex farming systems today is performed at different locations throughout Europe. The main aim is often to establish, adapt and verify efficient ("working") systems which may have a pilot character to change into "conventional" farming systems (IPM approach). Main criteria for their evaluation are economics and environmental effects. Some of the objectives are: (1) better economic returns, (2) reduced contamination of soil, water, air, agricultural products, humans, non-target organisms, (3) improved soil fertility, reduced soil erosion, (4) reduction in the abundance of pests and pathogens – improved "natural regulation", and (5) nature conservation.

Arthropod natural enemies in arable land · II *Survival, reproduction and enhancement*
C.J.H. Booij & L.J.M.F. den Nijs (eds.). *Acta Jutlandica* vol. 71:2 1996, pp. 269-275
 ISBN 87 7288 672 2

Important tools to achieve these goals are: (1) reduced pesticide input, (2) selective application of pesticides, (3) conservation tillage, (4) reduced fertilization, (5) resistant varieties, and (6) extended crop rotations.
In the discussion session the following aspects were treated in particular: conservation headlands, field boundaries, undersowing and mixed cropping and the increased diversity of the agricultural landscape.

Two main problems (drawbacks) from a scientific point of view were stressed:
– System approaches entail the difficulty of identifying and proving interrelationships between single and/or multiple factors, and causal connections often remain masked ("black box").
Frequently the experimental designs (replicates etc.) are inappropriate, especially when specific features of the involved organisms have to be considered, and when the demands of "sophisticated" statistic procedures have to be met.
– Convincing results can be expected only with long term investigations. Corrections and adjustments of the system based on significant results are sluggish and often delayed. The common practice of 3-year funding periods is unsuitable, particularly if more complex evaluations of ecosystem effects are regarded.

The common opinion during the discussion was that in principle it is very useful to continue with trials on farm scale using large experimental plots cultivated under practice-orientated farming conditions: The main objective is not primarily to elucidate the interdependence of single components but to develop an optimally adjusted functional package of measures and to test its efficiency. In the first place efficiency should not be measured by economic returns but by the parameters mentioned above including general aspects of nature conservation and biodiversity.

Manipulation of habitat structure

The second larger complex of the discussion dealt with feasible or proved effects of manipulation of habitat structure with regard to:
– Increased species richness and abundance
– Enhanced efficacy of natural enemies (improved pest "regulation").
Unfortunately, both aims are often mixed and not critically discussed with clear differentiation.

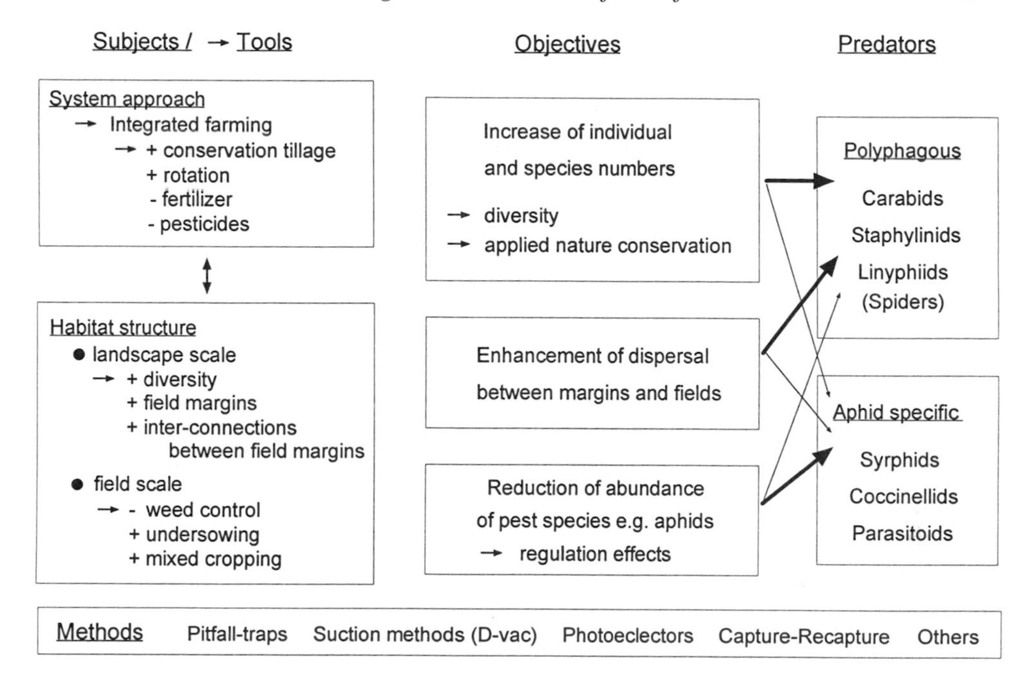

Fig. 1. Main subjects, objectives and predator groups in the field of enhancement of beneficial arthropods in farmland.

Applied nature conservation

Attempts to increase species richness and abundance of a broad spectrum of typical species in agro-ecosystems should not primarily focus (as often done) on the famous so-called "beneficials" but also take into consideration the often underrated "indifferent species". This can be valuable to:
– slow down or even stop the progressive loss of typical field species
– maintain a broad reservoir of genes (gene pool)
– increase the recreation value and aesthetics of our man-made landscape.

These aspects – here called "applied nature conservation" – are of fundamental importance today although there seem to be, superficially considered, no striking and immediate advantages for the economy of plant production systems. Never before during the recent years was there a better chance (or a better political climate) to incorporate constructive ideas for applied nature conservation in systems of sustainable integrated plant production.

The participants unanimously assessed that in addition to avoidance of direct negative effects (e.g. insecticides) increasing the diversity of agro-ecosystems can contribute to attain those objectives. Quite a range of valuable tools were mentioned, e.g. reduction of herbicide input, reduction of field sizes, creation of conservation headlands, recreation or establishment of boundaries with a heterogeneous vegetation, mixed cropping or even weedy strips within cultivated areas.

Habitat structure and the distribution (and efficacy?) of natural enemies

The second task: "What are convenient instruments to achieve a sounder pest control based on increased efficacy of natural enemies?" was debated with much more controversy. Often propagated are the following advantages of "diversity":

– Diversely structured field boundaries provide a wide range of prey/host resources, alternative food supply (e.g. pollen), shelter and a favourable microclimate for a wide range of field inhabiting polyphagous predators (carabids, staphylinids, spiders), specialized predators (syrphids, coccinellids) or parasitoids, temporarily living in the field.

– Diversity within fields by reduced weed control and/or undersowing or at least more or less mixed cropping systems may have comparable effects.

– Diversified rotations including management of set-aside areas may lead to a complex patchwork of differently structured habitats favouring exchange and migration on a larger scale.

Distribution of "mobile" species

The papers indicating better "pest control" in diversified areas are numerous. But is the fundamental role of the often praised antagonists, particularly of some polyphagous species really so significant in this context? It was concluded from the discussion that positive effects of field boundaries and a diversified landscape in general are obvious if mobile species are concerned. This is mainly relevant for aphid specific predators (Syrphids, Coccinellids), some parasitoids but also for some of the polyphagous species. All these mobile species are favoured in population build-up and synchronisation with target pests (e.g. aphids) in time and space (one important aspect of possible "efficacy" – see below) if refuges are available to sustain and compensate for periods of low food supply and unfavourable microclimatic conditions.

Distribution of polyphagous predators, the problem of persistent habitat preferences

However, there are increasing numbers of reports that for a wide range of polyphagous predators dispersal rates and distances between margins and cultivated areas are rather low. One major reason seems to be that a lot of species avoid sudden drops in habitat quality. Why should they leave the nice boundary, the "land of milk and honey", and explore the nasty surroundings of a wheat field? This, of course, is simplistic, however with a grain of truth, and with strong consequences for a better management of predator distribution. Although valuable data about migration processes, interchanges between fields and margins and the amount (intensity, frequency) of crossing of ecotones by different species are still rather scarce (it is not relevant here how far and straight the champion Pterostichus ... is able to run within 24 hrs), the discussion members felt that we have to put much more efforts in extending the convenient habitat qualities within the fields (perhaps smooth gradients) if a better distribution (and regulation ??) should be achieved. Some instruments may be:

– weedy ("dirty") fields
– reduced tillage, mulching
– inter- and mixed cropping

Future studies of dispersal in relation to influences on the population dynamics of key-pests are necessary.

General diversity to combat fluctuating pests?

Another short but interesting discussion point may be characterized by the following question:

"Is it convenient to fight against pest problems in unbalanced systems such as crops in annual rotations with long term increases in overall (general) diversity?"

This question cannot be answered sufficiently today by existing data (need for research efforts!). However, it may be useful and necessary to discuss it more in detail. If pests adapted to unbalanced systems can be efficiently controlled by biological agents only on a short term run with temporary operating (e.g. inundative released) antagonists ("task force police"), then all efforts to manage habitats are useless considering this special aspect.

Basic life-cycles, ecology of natural enemies and how "beneficial" are predators really?

A crucial drawback today for the above discussed problems and for suggested research needs is our limited knowledge of basic life-cycles and important ecological aspects of a lot of natural enemies, despite numerous more or less valuable studies. Although this is a really tedious work, it has to be done! In addition to our unsatisfactory knowledge about life-cycles of some predators and parasitoids it seems that we often use the rating "beneficial" too rashly. The fact alone that some predators show similar spatial and temporal distributions as the "desired" prey and that their density can be increased by some of the tools discussed above is not a reliable evidence for their predatory efficacy or their regulating potential. For a lot of predator groups, carabids as well as staphylinids or even spiders, we still speculate about their real effects on pest populations. Is there really a "regulation" in beneficial rich areas and to what extent? After passed enthusiastic periods about fantastic regulation capacities of different key beneficials from the polyphagous family as well as from the aphid-specific ones, more and more papers critically discuss the predatory effects, particularly if short-term (see above) pest control on low levels in annual crops is desired. This again offers a broad field for future research activities. Coming studies do not only have to intensify measurements of actual food consumption and prey preferences under field conditions but also the often neglected effects of inter- and intraspecific competition of beneficials. Particularly in studies of predator population dynamics and ecology and for the quantification of prey, predator interactions a large demand for models is obvious.

Conclusion – Summary

The discussion reflected that most attempts to increase the diversity of agro-ecosystems have some positive effects for "applied nature conservation" and that this is, independent of pest regulation, an important aspect of modern IPM-systems; however, more questions than answers arose concerning the role of natural enemies, particularly when their efficacy in relation to habitat management practices and function is to be emphasized. Based on much more basic studies on population dynamics and ecology the role of predators as "beneficials" in unbalanced systems like annual crops has to be more critically assessed.

Acknowledgement

All participants from the discussion group: J. Holland, U. Krause, G. Bujaki, B. Kromp, W. Büchs, M. Paoletti, S. Finch, G. de Snoo, R. Daamen and J. Noorlander, are thanked for their valuable contribution.

List of participants

Bilde, T. Institute of Biological Sciences, Department of Zoology, University of Aarhus, Bldg. 135, DK-8000 Aarhus C, Denmark

Booij, C.J.H. Research Institute for Plant Protection, Binnenhaven 5, P.O. Box 9060, NL-6700 GW Wageningen, The Netherlands

Bujaki, G.I. Department of Plant Protection, Gödöllö University, Påte K.U.I., 2100 Gödöllö, Hungary

Büchs, W. Biologisches Bundesanstalt für Land- und Forstwirtschaft, Institüt für Pflanzenschutz, Messeweg 11/12, D-38104 Braunschweig, Germany

Daamen, R. Research Institute for Plant Protection, Binnenhaven 5, P.O. Box 9060, NL-6700 GW Wageningen, The Netherlands

Dinter, A. Institut für Pflanzenpathologie und Pflanzenschutz der Universität Hannover, Henenhäusenstr. 2, D-30419 Hannover, Germany

Dijk, Th.S. van. Biological Station Wijster, Kampsweg 27, NL-9418 PD Wijster, The Netherlands

Finch, S, Institute of Horticultural Research, Wellesbourne, Warwick CV35 9EF, United Kingdom

Heimbach, U. Biologische Bundesanstalt fur Land- und Forstwirtschaft, Messeweg 11/12, D-38104 Braunschweig, Germany

Helenius, J. Institute of Crop Protection, Agricultural Research Centre, FIN-31600, Jokioinen, Finland

Holland, J. School of Biological Sciences, Dept. of Biology/Biomedical Science Building, Basett Crescent East, Southampton SO9 3TU, United Kingdom

Kopp, A. Institut für Pflanzenpathologie und Pflanzenschutz der Universität Hannover, Henenhäusenstr. 2, D-30419 Hannover, Germany

Krause, U. Institut für Pflanzenpathologie, Grisebachstr. 6, D-37077 Göttingen, Germany

Kromp, B. Ludwig Boltzmann Institut für Biologische Landbau, Rinnbockstr. 15, A-1110 Wien, Austria

Lemke, A. Institut für Pflanzenpathologie und Pflanzenschutz der Universität Hannover, Henenhäusenstr. 2, D-30419 Hannover, Germany

Lock, C.A.M. Research Institute for Plant Protection, P.O. Box 9060, NL-6700 GW Wageningen, The Netherlands

Lys, J.A. Zoological Institute, University of Bern, Baltzerstr. 3, CH-3012 Bern, Switzerland

Nijs, L.J.M.F. den. Research Institute for Plant Protection, P.O. Box 9060, NL-6700 GW Wageningen, The Netherlands

Noorlander, J. Research Institute for Plant Protection, P.O. Box 9060, NL-6700 GW Wageningen, The Netherlands

Paoletti, M.G. Universita degli stui di Padova, Dipartimento di Biologia, Via Trieste 75, 35121 Padova, Italy

Poehling, H.M. Institut für Pflanzenpathologie und Pflanzenschutz der Universität Hannover, Henenhäusenstr. 2, D-30419 Hannover, Germany

Powell, W. Rothamsted Experimental Station, Department of Entomology and Nematology, Harpenden, Herts. Al5 2JQ, United Kingdom

Snoo, G.R. de. Centre of Environmental Science, Leiden University, P.O. Box 9518, NL-2300 RA Leiden, The Netherlands

Sunderland, K.D. Horticultural Research International, Littlehampton, West Sussex BN17 6LP, United Kingdom

Szyszko, J. Intitute of Forest Protection and Ecology, Warszaw Agricultural University, Rakowiecka 26/30, Warszaw PL-02-528, Poland

Toft, S. Institute of Biological Sciences, Department of Zoology, University of Aarhus, Bldg. 135, DK-8000 Aarhus C, Denmark

Topping, C. Danish Environmental Institute, Kalo, Grenaavej 12, DK-8410 Ronde, Denmark

Zimmermann, J. Jonasstr. 21, D-12053, Berlin, Germany